CITES Orchid Checklist

Volume 4

For the genera:

Aerides, Coelogyne, Comparettia and *Masdevallia*

Compiled by:

Matthew J Smith, Chris Brodie, Jenny Kowalczyk, Sabina Michnowicz, H Noel McGough & Jacqueline A Roberts

Assisted by a selected international panel of orchid experts

Royal Botanic Gardens, Kew

© Copyright 2006
The Trustees of The Royal Botanic Gardens, Kew

First published in 2006

General editor of series: Jacqueline A Roberts

ISBN 1 84246 122 2

Produced with the financial assistance of
the CITES Nomenclature Committee and
the Royal Botanic Gardens, Kew

For information or to purchase Kew titles please visit
www.kewbooks.com or email publishing@kew.org

Cover design by Media Resources RBG Kew

Acknowledgements / Remerciements / Agradecimientos

The compilers would like to thank colleagues at the Royal Botanic Gardens, Kew and the following orchid experts for their help with the preparation of the checklist for publication. All suggestions for amendments were gratefully noted and included at the discretion of the compilers. We would particularly like to thank **Andre Schuiteman, Ed de Vogel** and **Johan Hermans** for their comprehensive review of the final version of the text.

Les compilateurs tiennent à remercier leurs collègues des Royal Botanic Gardens de Kew et les experts en orchidées suivants, pour leur aide dans la préparation de la Liste en vue de sa publication. Il a été pris note avec gratitude de toutes les suggestions, qui ont été incluses à la discrétion des compilateurs. Nous adressons des remerciements particuliers à **Andre Schuiteman**, **Ed de Vogel** et **Johan Hermans** qui ont examiné de manière approfondie la version finale du texte.

Las personas encargadas de la recopilación de este volumen desean dar las gracias a sus colegas del Royal Botanic Gardens, Kew, así como los siguientes especialistas en orquídeas por su ayuda en la preparación de la presente lista. Se tomó nota con agradecimiento de todas las sugerencias de enmienda, que fueron incluidas conforme a los criterios de los responsables de este volumen. En particular deseamos dar las gracias a **Andre Schuiteman**, **Ed de Vogel** y **Johan Hermans** por la revisión de la versión definitiva del texto.

International Panel of Orchid Experts
Groupe international d'spécialistes des orchidées
Grupo internacional de expertos en orquídeas

Dr L Averyanov	Russia
Dr I Bock	Germany
Dr H J Chowdhery	India
Mr D Clayton	UK
Mrs I F la Croix	UK
Dr A Kocyan	Germany
Dr S Kumar	India
Mr E S Manning	UK
Mr G McDonald	South Africa
Mr W Rhodehamel	USA
Dr G Romero	USA
Mr T Sijm	the Netherlands
Mr D Smallman	UK
Dr M Wolff	Germany

CONTENTS

Preamble

TABLE DES MATIERES

Préambule

ÍNDICE

Preámbulo

CITES CHECKLIST – ORCHIDACEAE

PREAMBLE

1. Background

The 1992 Conference of the Parties to the Convention on International Trade in Endangered Species of Wild Fauna and Flora (CITES) adopted Resolution Conf. 8.19 which called for the production of a standard reference to the names of Orchidaceae.

The Vice-Chairman of the CITES Nomenclature Committee was charged with the responsibility of co-ordinating the input needed to produce such a reference.

The orchid genera identified as priorities in the *Review of Significant Trade in Species of Plants included in Appendix II of CITES* (CITES Doc. 8.31) would be treated first. The checklists (or parts thereof) as they came available would be put to the Conference of the Parties for approval.

At its third meeting (Chiang Mai, Thailand, November 1992) the Plants Committee extensively discussed a proposal by the Vice-Chairman of the Nomenclature Committee regarding the possible mechanisms to develop the Standard Reference. The Plants Committee endorsed a procedure by which compilations from the available literature, made on a central database, would be circulated to a panel of international experts for consultation and final decisions on the valid names to be used for the taxa concerned. During the development of the checklists every effort was made to recruit national experts from the range states using the contact network of the regional representatives of the Plants Committee. The response was poor and it is hoped that publication of the checklists will encourage increased participation in future volumes.

2. Computer aspects

Hardware: The database was set up on a Networked Pentium III PC using ALICE software, version 2.1.

Database system: The ALICE database system was used to handle the data collection. ALICE handles distribution, uses, common names, descriptions, habitats, synonymy, bibliography and other classes of data for species, subspecies or varieties. It allows users to design, for example, their own reports for checklists, studies, monographs or conservation lists.

ALICE software can be contacted by e-mail at info@alicesoftware.com
Webpage http://www.alicesoftware.com

3. Compilation procedures

- Primary references were identified by orchid specialists based at the Royal Botanic Gardens, Kew.
- An International Panel of Orchid Experts was established in order to review each stage of the checklist.
- Information was entered into the ALICE taxonomic database and a preliminary report produced.
- Preliminary reports for each genus were distributed to the Panel of Orchid Experts for their comments, additions or amendments.
- Additions and amendments returned from the Panel members were entered into the database. These were linked to a reference contained in the bibliography at the end of the report on each genus. (Not included in this checklist, but copies held for reference at RBG, Kew).

Preamble

- This sequence was repeated five times for each genus to allow full consultation with the Panel.
- A 'final draft' including all the genera was prepared and distributed to further experts for their comments
- Andre Schuiteman, Ed de Vogel and Johan Hermans were consulted as external editors to review the final draft.
- Any further additions and amendments were added to the database.
- Format for publication was agreed with the CITES Secretariat and reports generated using AWRITE and prepared for camera-ready copy using Microsoft Word for Windows 2000.

4. Conservation

During the consultation process with the Panel of Orchid Experts, information was also requested on the conservation status of the species concerned. Copies of the present and proposed new IUCN categories of threat were distributed to the Panel for their use.

5. How to use the Checklist

It is intended that this Checklist be used as a quick reference for checking accepted names, synonymy and distribution. The reference is therefore divided into three main parts:

Part I: All names in current use

An alphabetical list of all accepted names and synonyms included in this checklist - a total of 1738 names (771 accepted and 967 synonyms).

Part II: Accepted names in current use

Separate lists for each genus. Each list is ordered alphabetically by the accepted name and details are given on current synonyms and distribution.

Part III: Country checklist

Accepted names from all genera included in this Checklist are ordered alphabetically under country of distribution.

6. Conventions employed in Parts I, II and III

    a)   Accepted names are presented in **bold roman** type.
           Synonyms are presented in *italic* type.

    b)   Duplicate names

           In Part I, the Author's name appears after each taxon where the taxon name appears twice or more e.g., *Aerides angustifolia* Hook.f., *Aerides angustifolia* (Blume) Lindl. non Hook.f. (unless the author's name is the same).

    i)   Where a synonym occurs more than once, but refers to different species, for example, *Aerides latifolium*, for both **Aerides odorata** and **Phalaenopsis deliciosa**, an asterisk indicates the species most likely to be encountered in trade - if this is known. For example:

All Names	Accepted Name
Aerides latifolium (Thunb. ex Sw.) Sw.	**Aerides odorata***
Aerides latifolium Thwaites non (Thunb. ex Sw.) Sw	**Phalaenopsis deliciosa**

           *Species most likely to be in trade (in this example, **Aerides odorata**).

2

ii) Where an accepted name and a synonym are the same, but refer to different species, for example, **Masdevallia meleagris** (accepted name) and *Masdevallia meleagris* (a synonym of **Masdevallia picturata**), the name with an asterisk is the species most likely to be seen in trade - if this is known. For example:

All Names   **Accepted Name**

Masdevallia meleagris Lindl.*
Masdevallia meleagris Lindl. sensu Rchb.f. non Lindl. ... **Masdevallia picturata**

*Species most likely to be in trade (in this example, **Masdevallia meleagris**).

NB: In examples bi) and bii) it is necessary to double-check by reference to the distribution as detailed in Part II. For instance, in the example bii), if the name given was '*Masdevallia meleagris*' and it was known that the plant in question came from Guyana this would indicate that the species was **Masdevallia picturata,** being traded under the synonym *Masdevallia meleagris*. **Masdevallia meleagris** is only found in Colombia.

c) Natural hybrids have been included in the checklist and are indicated by the multiplication sign ×. They are arranged alphabetically within the lists.

d) Many *Aerides* species epithets have been published with an incorrect Latin termination. Where appropriate, species epithets should have a Latin feminine termination, as this is how the generic name was originally published. However, the majority of names have been published with neuter species epithet terminations. These species epithets have been changed to the correct feminine terminations. This is in accordance with the International Code of Botanical Nomenclature, (2000).

As the neuter epithets are in common use and used in the CITES trade data, primary references and other major references, we have included the neuter epithets as synonyms.

e) The CD-ROM contains the following files:

'CITESOrchidChecklist4.pdf', an Adobe Acrobat®. This file contains CITES Orchid Checklist 4 and can be viewed using Adobe Reader®. You will need Adobe Acrobat Reader® installed on your computer to view this file (can be downloaded from www.adobe.com).

Navigation of the Adobe Acrobat® file is made easy using either the Thumbnails on the left hand side of the document or active links within the document. All active links within the document are highlighted in blue.

'OrchidIntro.pdf', an Adobe Acrobat®. This files contains the introductory text to the CITES Orchid Checklists Volumes 1-3. The file can be viewed using Adobe Reader®. You will need Adobe

Preamble

Acrobat Reader® installed on your computer to view this file (can be downloaded from www.adobe.com).

'Vols I II and III combined_1.pdf', an Adobe Acrobat®. This files contains an alphabetical list of all accepted names and synonyms for the genera as included in CITES Orchid Checklists Volumes 1-3. The file can be viewed using Adobe Reader®. You will need Adobe Acrobat Reader® installed on your computer to view this file (can be downloaded from www.adobe.com).

'Vols I II and III combined_2.pdf', an Adobe Acrobat®. This files contains an ordered alphabetical list of the accepted names and their current synonyms and distribution for the genera in CITES Orchid Checklists Volumes 1-3. The file can be viewed using Adobe Reader®. You will need Adobe Acrobat Reader® installed on your computer to view this file (can be downloaded from www.adobe.com).

'Vols I II and III combined_3.pdf', an Adobe Acrobat®. This files contains Accepted names from all genera included in CITES Orchid Checklists Volumes 1-3, ordered alphabetically under country of distribution. The file can be viewed using Adobe Reader®. You will need Adobe Acrobat Reader® installed on your computer to view this file (can be downloaded from www.adobe.com).

7. Number of names entered for each genus:
Aerides (Accepted: 23, Synonyms: 290); *Coelogyne* (Accepted: 194, Synonyms: 287); *Comparettia* (Accepted: 6, Synonyms: 7); *Masdevallia* (Accepted: 548, Synonyms: 383).

8. Geographical areas
Country names follow the United Nations standard as laid down in Country Names. *Terminology Bulletin* August 1997. United Nations 347:1-41.

9. Orchidaceae controlled by CITES
The family Orchidaceae is listed on Appendix II of CITES. In addition the following taxa are listed on Appendix I at time of publication:

Aerangis ellisii
Dendrobium cruentum
Laelia jongheana
Laelia lobata
Paphiopedilum spp.
Peristeria elata
Phragmipedium spp.
Renanthera imschootiana

10. Abbreviations, botanical terms and Latin*
Not all these abbreviations, botanical terms and Latin will appear in this Checklist; however, they have been included as a useful reference. Note: words in *italics* are Latin.

ambiguous name a name which has been applied to different taxa by different authors, so that it has become a source of ambiguity
anon. anonymous; without author or author unknown
auct. *auctorum*: of authors

CITES Convention on International Trade in Endangered Species of Wild Fauna and Flora

cultivar an individual, or assemblage of plants maintaining the same distinguishing features, which has been produced or is maintained (propagated) in cultivation

cultivation the raising of plants by horticulture or gardening; not immediately taken from the wild

descr. *descriptio*: the description of a species or other taxonomic unit

distribution where plants are found (geographical)

ed. editor

edn. edition (book or journal)

eds. editors

epithet the last word of a species, subspecies, or variety (etc.), for example: *speciosa* is the species epithet for the species *Coelogyne speciosa*.

escape a plant that has left the boundaries of cultivation (e.g. a garden) and is found occurring in natural vegetation

ex *ex*: after; may be used between the name of two authors, the second of whom validly published the name indicated or suggested by the first

excl. *exclusus*: excluded

forma *forma*: a taxonomic unit inferior to variety

hort. *hortorum*: of gardens (horticulture); raised or found in gardens; not a plant of the wild

ICBN International Code for Botanical Nomenclature

in prep. in preparation

in sched. *in scheda*: on a herbarium specimen or label

in syn. *in synonymia*: in synonymy

incl. including

ined. *ineditus*: unpublished

introduction a plant which occurs in a country, or any other locality, due to human influence (by purpose or chance); any plant which is not native

key a written system used for the identification of organisms (e.g. plants)

leg. *legit*: he gathered; the collector

misspelling a name that has been incorrectly spelt; not a new or different name

morphology the form and structure of an organism (e.g. a plant)

name causing confusion a name that is not used because it cannot be assigned unambiguously to a particular taxon (e.g. a species of plant)

native an organism (e.g. a plant) that occurs naturally in a country, or region, etc.

naturalized a plant which has either been introduced (see introduction) or has escaped (see escape) but which looks like a wild plant and is capable of reproduction in its new environment

nec *neque*: and not, also not, neither

nom. *nomen*: name

nom. ambig. *nomen ambiguum*: ambiguous name

nom. cons. prop. *nomen conservandum propositum*: name proposed for conservation under the rules of the International Code for Botanical Nomenclature (ICBN)

nom. illeg *nomen illegitimum*: illegitimate name

nom. nud. *nomen nudum*: name published without description

nomenclature branch of science concerned with the naming of organisms (e.g. plants)

non *non*: not

only known from cultivation a plant which does not occur in the wild, only in cultivation

Preamble

orthographic variant an alternative spelling for the same name
p.p. *pro parte*: partly, in part
provisional name name given in anticipation of a valid description
sens. *sensu*: in the sense of; the manner in which an author interpreted or used a name
sens. lat. *sensu lato*: in the broad sense; a taxon (usually a species) and all its subordinate taxa (e.g. subspecies) and/or other taxa sometimes considered as distinct
sensu *sensu*: in the sense of; the manner in which an author interpreted or used a name
sic *sic*, used after a word that looks wrong or absurd, to show that it has been quoted correctly
spp. species
ssp. subspecies
synonym a name that is applied to a taxon but which cannot be used because it is not the accepted name – the synonym or synonyms form the synonymy
taxa plural of taxon
taxon a named unit of classification, e.g. genus, species, subspecies
var. variety

*thanks to Dr Aaron Davis, RBG Kew, for the provision of this guide.

11. Bibliography

Primary reference sources used in the compilation of checklists:

Averyanov, L.V. & Averyanova, A.L. (2003). *Updated checklist of the orchids of Vietnam*. Vietnam National University Publishing House. Hanoi.

Bock, I. (1986). Revision der Gattung *Comparettia* Poepp. & Endl. (Teil 1) *Die Orchidee*. 37 (4): 192-196.

Bock, I. (1986). Revision der Gattung *Comparettia* Poepp. & Endl. (Teil 2). *Die Orchidee*. 37(5): 199-196.

Bock, I. (1986). Revision der Gattung *Comparettia* Poepp. & Endl. (Teil 3). *Die Orchidee*. 37(6): 255-263.

Brummit, R.K. & Powell C.E. (1992). *Authors of plant names*. Royal Botanic Gardens, Kew. UK.

Chen, S. [*et al.*]: (1999). *Angiospermae Flora reipublicae popularis Sinicae tomus 18. Monocotyledoneae*. Orchidaceae 2. Science Press, China.

Christenson, E.A. (1987). The taxonomy of Aerides & related genera in: Kamezo, S. & Tanaka, R. (eds), (1987). *Proceedings of the 12th World Orchid Conference*, World Orchid Conference, Tokyo, 12: 35-40.

Clayton, D. (2002). *The genus Coelogyne: a synopsis*. Royal Botanic Gardens, Kew. UK.

Comber, J.B. (2001). *Orchids of Sumatra*. Royal Botanic Gardens, Kew. UK.

Gagnepain, F. & Guillaumin, A. (1932 - 1934). *Orchidacees*. In Gagnepain, F (ed) (1908 – 1942). Flore Generale de L'Indo-Chine. 16: 142-647.

Cribb, P.J. & Wood, J.J. (1994). *A Checklist of the Orchids of Borneo.* Royal Botanic Gardens, Kew, UK.

Kraenzlin, F.W.L. (1925). *Repertorium Speciecum nobarum regni vegetabilis, Monographie der Guttungen Masdevallia.* Dahlem bei Berlin. Germany.

Luer, C.A. (1983 - 1998). *Thesaurus Masdevalliarum: a monograph of the genus Masdevallia.* Volumes 1-20(a), Missouri Botanical Garden, USA.

Luer, C.A. (1998 - 2001). *A treasure of Masdevallia: a monograph of the genus Masdevallia.* Volumes 21-26, Missouri Botanical Garden, USA.

Luer, C.A. (2000). *Icones Pleurothallidinarum XIX. Systematics of Masdevallia: Part one.* Missouri Botanical Garden, USA.

Luer, C.A. (2000). *Icones Pleurothallidinarum XXI. Systematics of Masdevallia: Part Two.* Missouri Botanical Garden, USA.

Luer, C.A. (2001). *Icones Pleurothallidinarum XXII. Systematics of Masdevallia: Part Three.* Missouri Botanical Garden, USA.

Luer, C.A. (2002). *Icones Pleurothallidinarum XXIII. Systematics of Masdevallia: Part Four.* Missouri Botanical Garden, USA.

Luer, C.A. (2003). *Icones Pleurothallidinarum XXV. Systematics of Masdevallia: Part Five.* Missouri Botanical Garden, USA.

Ormerod, P. (1997). A Review of Coelogyne sect. Proliferae. *Austr. Orch. Rev.* 19-23.

Pearce, N.R. & Cribb, P.J. (2002). *The Orchids of Bhutan.* Royal Botanic Garden, Edinburgh. UK.

Seidenfaden, G. (1977/78). Orchid genera in Thailand: I-III. *Dansk Botanisk Arkiv* 29: 3-4.

Seidenfaden, G. (1988). Orchid genera in Thailand: XIV. Fifty-nine vandoid genera. *Opera Botanicam,* Vol 95.

Seidenfaden, G. (1992). The Orchids of Indochina. *Opera Botanica,* Vol 114.

Seidenfaden, G. & Wood, J.J. (1992). *The Orchids of Peninsular Malaysia and Singapore; a revision of R. E. Holttum: Orchids of Malaya*, Olsen & Olsen, Fredensborg, Denmark.

Stern, W.T. (1992). *Botanical Latin (fourth edition).* David & Charles Publishers, Newton Abbot, Devon. UK.

The International Plant Names Index (2004). Published on the internet, http://www.ipni.org.

World Checklist of Monocots (2004). The Board of Trustees of the Royal Botanic Gardens, Kew. Published on the internet, www.kew.org/monocotChecklist/.

LISTE CITES DES ORCHIDACEAE

PRÉAMBULE

1. Contexte

En 1992, la Conférence des Parties à la Convention sur le commerce international des espèces de faune et de flore sauvages menacées d'extinction (CITES) a adopté la résolution Conf. 8.19 dans laquelle elle recommande la préparation d'une liste normalisée de référence des noms d'Orchidaceae.

Le vice-président du Comité CITES de la nomenclature a été chargé de coordonner les informations reçues en vue de préparer cette liste.

Les genres d'orchidées classés comme prioritaires dans l'Examen du commerce important d'espèces végétales inscrites à l'Annexe II de la CITES (document CITES Doc. 8.31) devaient être les premiers traités. Les listes (ou parties de listes) devaient être soumises à la Conférence des Parties pour approbation à mesure qu'elles seraient disponibles.

A sa troisième session (Chiang Mai, Thaïlande, novembre 1992) le Comité pour les plantes a abondamment discuté d'une proposition du vice-président du Comité de la nomenclature concernant les mécanismes possibles d'élaboration d'une liste de référence normalisée. Le Comité pour les plantes a approuvé une procédure par laquelle les compilations faites à partir de la littérature disponible, sur une base de données centrale, seraient envoyées à un groupe d'experts internationaux pour consultation et décision finale sur les noms valides devant être utilisés pour les taxons concernés.

Au cours de la préparation des listes, il a été fait appel aux spécialistes nationaux des Etats de l'aire de répartition, en utilisant le réseau de contacts des représentants régionaux du Comité pour les plantes. Toutefois, il y a eu peu de réponses. L'on espère que la publication de la présente Liste favorisera une participation accrue aux futurs volumes.

2. Aspects informatiques

Matériel: La base de données a été créée sur un PC Pentium III en UTILISANT le logiciel ALICE version 2.1.

Système de base de données: Le système ALICE a été utilisé pour enregistrer les données. ALICE traite la répartition géographique, les utilisations, les noms communs, les descriptions, les habitats, les synonymes, la bibliographie et d'autres catégories de données relatives aux espèces, sous-espèces ou variétés. Il permet aux utilisateurs de créer leurs propres rapports sous forme de listes, d'études, de monographies, de listes de conservation, etc.

ALICE Software peut être contacté par courriel à: info@alicesoftware.com
Sur Internet: http://www.alicesoftware.com

3. Procédure de compilation

- Les principales références ont été identifiées par les spécialistes des orchidées des Royal Botanic Gardens, Kew.
- Un groupe international de spécialistes des orchidées a été établi afin d'examiner la Liste à chaque étape.

Préambule

- Des informations ont été introduites dans la base de données taxonomiques ALICE et un rapport préliminaire a été préparé.
- Des rapports préliminaires sur chaque genre ont été remis au groupe de spécialistes pour qu'ils formulent ses commentaires sur les additions ou amendements nécessaires.
- Les ajouts et amendements des membres du groupe ont été introduits dans la base de données. Ils ont été reliés à une référence contenue dans la bibliographie à la fin du rapport sur chaque genre. (Non comprise dans la présente Liste mais dont des copies sont à disposition, pour servir de référence, auprès de RBG, Kew).
- Le processus a été répété cinq fois pour chaque genre afin de permettre la pleine consultation du groupe.
- Un « project final » couvrant tous les genres a été préparé et envoyé à d'autres spécialistes en leur demandant leurs commentaires.
- Andre Schuiteman, Ed de Vogel et Johan Hermans ont été consultés en tant que réviseurs externes et ont examiné le projet final.
- Tous les autres autres ajouts et amendements ont été ajoutés à la base de données.
- La présentation retenue pour la publication a été convenue avec le Secrétariat CITES et les rapports ont été créés en utilisant AWRITE et préparés pour la publication en utilisant Microsoft Word pour Windows, 2000.

4. Conservation

Au cours du processus de consultation du groupe d'experts, des informations ont été demandées sur l'état de conservation des espèces concernées. Le groupe a reçu des copies des catégories actuelles et des nouvelles catégories de menaces proposées par l'UICN.

5. Comment utiliser la Liste?

Cette Liste devrait être utilisée comme liste de référence pour vérifier rapidement les noms acceptés, les synonymes et la répartition géographique. Elle est divisée en trois parties principales:

Première partie : binomes d'orchidaceae acuellement en usage
Liste alphabétique de tous les noms acceptés et des synonymes inclus dans la Liste : 1738 noms (771 noms acceptés et 967 synonymes).

Deuxième partie : noms acceptés d'usage courant
Il existe une liste pour chaque genre. Chaque liste contient les noms acceptés par ordre alphabétique et comporte des indications sur les synonymes et la répartition géographique actuels.

Troisième partie : liste par pays
Les noms acceptés de tous les genres inclus dans cette Liste apparaissent par ordre alphabétique pour chaque pays de l'aire de répartition.

6. Conventions utilisées dans la première, la deuxième et la troisième partie
Les noms acceptés sont en **caractères gras.**
Les synonymes sont en *italique.*

b)   Noms identiques pour des taxons différents :

Dans la première partie, le nom de l'auteur apparaît après chaque taxon lorsque le taxon est mentionné deux fois ou plus; par exemple : *Aerides angustifolia* Hook.f. et *Aerides angustifolia* (Blume) Lindl. non Hook.f. (sauf si le nom de l'auteur est le même).

i) Lorsque le synonyme apparaît plus d'une fois mais fait référence à des noms acceptés différents – par exemple, *Aerides latifolium*, (synonyme à la fois de **Aerides odorata** et de **Phalaenopsis deliciosa**), le nom comportant un astérisque est celui de l'espèce la plus susceptible d'être trouvée dans le commerce, si on le connaît. Exemple:

Tous les noms	Nom acceptés
Aerides latifolium (Thunb. ex Sw.) Sw.	**Aerides odorata***
Aerides latifolium Thwaites non (Thunb. ex Sw.) Sw...	**Phalaenopsis deliciosa**

*Espèce la plus susceptible d'être trouvée dans le commerce (dans cet exemple, **Aerides odorata**)

ii) Lorsque le nom accepté et un synonyme sont les mêmes mais font référence à des espèces différentes - par exemple **Masdevallia meleagris** (nom accepté) et *Masdevallia meleagris* (synonyme de **Masdevallia picturata**), le nom comportant un astérisque est celui de l'espèce la plus susceptible d'être trouvée dans le commerce, si on le connaît. Exemple:

Tous les noms	Nom acceptés
Masdevallia meleagris Lindl.*	
Masdevallia meleagris Lindl. sensu Rchb.f. non Lindl. ...	**Masdevallia picturata**

*Espèce la plus susceptible d'être trouvée dans le commerce (dans cet exemple, **Masdevallia meleagris**)

NB : Dans les exemples b i) et b ii), il faut effectuer une double vérification en se référant à la répartition géographique indiquée dans la deuxième partie. Ainsi, dans l'exemple b ii), si le nom donné est *Masdevallia meleagris* et si l'on sait que la plante vient de Guyane, cela indique qu'il s'agit de **Masdevallia picturata** commercialisée sous le synonyme *Masdevallia meleagris*. **Masdevallia meleagris** ne pousse qu'en Colombie.

c) Les hybrides naturels figurent dans les listes par ordre alphabétique et sont indiqués par le signe de multiplication ×.

d) Bon nombre d'épithètes d'espèces du genre *Aerides* ont été publiés avec une terminaison incorrecte en Latin. Dans certains cas, les épithètes des espèces devraient avoir une terminaison féminine en Latin, car c'est ainsi que le nom générique a été publié à l'origine. Cependant, la plupart des noms ont été publiés avec des terminaisons neutres pour les épithètes des espèces. Ces épithètes ont été corrigés et remplacés par les terminaisons féminines correctes, conformément au Code international de la nomenclature botanique (2000).

Préambule

Etant donné que les épithètes neutres sont d'usage courant et apparaissent dans les informations sur le commerce CITES, les références principales et autres références importantes, nous avons ajouté les épithètes neutres en tant que synonymes.

e)   Le CD-ROM contient les fichiers suivants:

'CITESOrchidChecklist4.pdf', un fichier Adobe Acrobat® qui contient le livre CITES Orchid Checklist 4 et peut être visualisé en utilisant Adobe Reader®. Vous aurez besoin de Adobe Acrobat Reader® pour visualiser ce fichier. Pour télécharger ce logiciel, visitez www.adobe.com.

Les signets (« bookmarks ») à gauche de l'écran ou les liens actifs qui se trouvent dans le document lui-même permettent de naviguer aisément en utilisant Adobe Acrobat®. Tous les liens actifs du document sont affichés en bleu.

'OrchidIntro.pdf', un fichier Adobe Acrobat® qui contient l'introduction aux Volumes 1-3 des Listes des Orchidées CITES et peut être visualisé en utilisant Adobe Reader®. Vous aurez besoin de Adobe Acrobat Reader® pour visualiser ce fichier. Pour télécharger ce logiciel, visitez www.adobe.com.

'Vols I II and III combined_1.pdf', un fichier Adobe Acrobat® qui contient une liste par ordre alphabétique de tous les noms acceptés et les synonymes pour les genres, comme celles qui apparaissent dans les Volumes 1-3 des Listes des Orchidées CITES. Ce fichier peut être visualisé en utilisant Adobe Reader®. Vous aurez besoin de Adobe Acrobat Reader® pour visualiser ce fichier. Pour télécharger ce logiciel, visitez www.adobe.com.

'Vols I II and III combined_2.pdf', un fichier Adobe Acrobat® qui contient une liste par ordre alphabétique des noms acceptés et des synonymes actuels, ainsi que de la distribution du genre, comme celles qui apparaissent dans les Volumes 1-3 des Listes des Orchidées CITES. Ce fichier peut être visualisé en utilisant Adobe Reader®. Vous aurez besoin de Adobe Acrobat Reader® pour visualiser ce fichier. Pour télécharger ce logiciel, visitez www.adobe.com

'Vols I II and III combined_3.pdf', un fichier Adobe Acrobat® qui contient les noms acceptés de tous les genres compris dans les Volumes 1-3 des Listes des Orchidées CITES, par ordre alphabétique sous la rubrique « pays de distribution ». Ce fichier peut être visualisé en utilisant Adobe Reader®. Vous aurez besoin de Adobe Acrobat Reader® pour visualiser ce fichier. Pour télécharger ce logiciel, visitez www.adobe.com

7. Nombre de noms compris pour chaque genre:
Aerides (acceptés : 23, synonymes : 290); *Coelogyne* (acceptés : 194, synonymes : 287); *Comparettia* (acceptés : 6, synonymes 7); *Masdevallia* (acceptés : 548, synonymes : 383).

8. Régions géographiques

Les noms des pays sont ceux figurant dans le *Terminology Bulletin No. 347* des Nations Unies, août 1997, pages 1 - 41.

9. Orchidées soumises aux contrôles CITES

La famille des Orchidaceae est inscrite à l'Annexe II de la CITES. De plus, les taxons suivants étaient inscrits à l'Annexe I au moment de la publication de la Liste:

Aerangis ellisii
Dendrobium cruentum
Laelia jongheana
Laelia lobata
Paphiopedilum spp.
Peristeria elata
Phragmipedium spp.
Renanthera imschootiana

10. Abréviations, termes botaniques et mots latins *

Ces termes de botanique, noms latins et abréviations ne sont pas tous utilisés dans la Liste. Ils sont fournis pour servir de référence. Note : les mots *en italique* sont d'origine latine.

ambiguous name (nom ambigu) nom donné à différents taxons par différents auteurs, ce qui crée une ambiguïté

anon. anonyme; sans auteur

auct. *auctorum* : d'auteurs

CITES Convention sur le commerce international des espèces de faune et de flore sauvages menacées d'extinction

cultivar spécimen ou groupe de plantes conservant les mêmes caractéristiques distinctives, produites ou conservées (propagées) en culture

cultivation (culture) obtention de plantes par horticulture ou jardinage, par opposition au prélèvement dans la nature

descr. *descriptio* description d'une espèce ou d'une autre entité taxonomique

distribution (aire de répartition géographique) région(s) où se trouve les plantes

ed. éditeur

edn. édition (d'un livre ou d'un périodique)

eds. éditeurs

epithet (épithète) dernier mot d'une espèce, d'une sous-espèce ou d'une variété (etc.). Exemple : *speciosa* est l'épithète de l'espèce *Coelogyne speciosa.*

escape (échappée) qualifie une plante qui a quitté l'enceinte de culture (un jardin, par exemple) et qu'on retrouve dans la végétation naturelle

ex *ex* d'après; peut être utilisé entre deux noms d'auteurs, dont le second a validement publié le nom d'après les indications ou suggestions du premier

excl. *exclusus* exclu

hort. *hortorum* de jardins (horticole); plante cultivée ou se trouvant dans des jardins horticoles, par opposition à une plante d'origine sauvage

ICBN (CINB) Code international de la nomenclature botanique

in prep. en préparation

in sched. *in scheda* sur un spécimen d'herbier ou une étiquette

in syn. *in synonymia* en synonymie

incl. incluant

ined. *ineditus* non publié

introduction résultat d'une activité humaine (volontaire ou non) aboutissant à ce qu'une plante non indigène se retrouve dans un pays ou une région

key (clé) système écrit utilisé pour la détermination d'organismes (plantes, par exemple)

leg. *legit* il ramassa; le collecteur

misspelling (faute d'orthographe) nom mal orthographié, par opposition à un nom nouveau ou différent

morphology (morphologie) forme et structure d'un organisme (d'une plante, par exemple)

name causing confusion (nom causant une confusion) nom qui n'est pas utilisé parce qu'il ne peut être assigné sans ambiguïté à un taxon particulier (à une espèce de plante, par exemple)

native (indigène) qualifie un organisme (une plante, par exemple) prospérant naturellement dans un pays ou une région etc.

naturalized (naturalisée) qualifie une plante introduite (voir introduction) ou échappée (voir échappée) qui ressemble à une plante sauvage et qui se propage dans son nouvel environnement

nec *neque* : et non, et ... ne ... pas, ni, non plus, et non

nom. *nomen* nom

nom. ambig. *nomen ambiguum* nom ambigu

nom. cons. prop. *nomen conservandum propositum* nom dont le maintien a été proposé d'après les règles du *International Code of Botanical Nomenclature* (Code international de la nomenclature botanique)

nomenclature branche de la science qui nomme les organismes (les plantes, par exemple)

non *non* pas

only known from cultivation (connue seulement en culture) qualifie une plante qu'on ne trouve pas à l'état sauvage

orthographic variant (variante orthographique) même nom orthographié différemment

pro parte *pro parte* partiellement, en partie

provisional name (nom provisoire) nom donné par anticipation d'une description **sens.** *sensu* au sens de; manière dont un auteur interprète ou utilise un nom

sens. lat. *sensu lato* au sens large; un taxon (habituellement une espèce) et tous ses taxons inférieurs (sous-espèce, etc.) et/ou d'autres taxons parfois considérés comme distincts

sic *sic*, utilisé après un mot qui semble faux ou absurde; indique que ce mot est cité textuellement

synonym (synonyme) nom donné à un taxon mais qui ne peut être utilisé parce que ce n'est pas le nom accepté; le ou les synonymes forment la synonymie

taxa pluriel de taxon

taxon unité taxonomique à laquelle on a attribué un nom - genre, espèce, sous-espèce,
etc.

var. variété

* Nous remercions M. Aaron Davis, de RBG Kew, d'avoir fourni ce guide.

11. Bibliographie
Principales sources de références utilisées dans la compilation des listes:

Averyanov, L.V. & Averyanova, A.L. (2003). *Updated checklist of the orchids of Vietnam*. Vietnam National University Publishing House. Hanoi.

Bock, I. (1986). Revision der Gattung *Comparettia* Poepp. & Endl. (Teil 1*) Die Orchidee*. 37 (4): 192-196.

Bock, I. (1986). Revision der Gattung *Comparettia* Poepp. & Endl. (Teil 2). *Die Orchidee*. 37(5): 199-196.

Bock, I. (1986). Revision der Gattung *Comparettia* Poepp. & Endl. (Teil 3). *Die Orchidee*. 37(6): 255-263.

Brummit, R.K. & Powell C.E. (1992). *Authors of plant names*. Royal Botanic Gardens, Kew. UK.

Chen, S. [*et al.*]: (1999). *Angiospermae Flora reipublicae popularis Sinicae tomus 18. Monocotyledoneae*. Orchidaceae *2*. Science Press, China.

Christenson, E.A. (1987). The taxonomy of Aerides & related genera in: Kamezo, S. & Tanaka, R. (eds), (1987). *Proceedings of the 12th World Orchid Conference*, World Orchid Conference, Tokyo, 12: 35-40.

Clayton, D. (2002). *The genus Coelogyne: a synopsis*. Royal Botanic Gardens, Kew. UK.

Comber, J.B. (2001). *Orchids of Sumatra*. Royal Botanic Gardens, Kew. UK.

Gagnepain, F. & Guillaumin, A. (1932 - 1934). *Orchidacees*. In Gagnepain, F (ed) (1908 – 1942). Flore Generale de L'Indo-Chine. 16: 142-647.

Cribb, P.J. & Wood, J.J. (1994). *A Checklist of the Orchids of Borneo*. Royal Botanic Gardens, Kew, UK.

Kraenzlin, F.W.L. (1925). *Repertorium Speciecum nobarum regni vegetabilis, Monographie der Guttungen Masdevallia*. Dahlem bei Berlin. Germany.

Luer, C.A. (1983 - 1998). *Thesaurus Masdevalliarum: a monograph of the genus Masdevallia*. Volumes 1-20(a), Missouri Botanical Garden, USA.

Luer, C.A. (1998 - 2001). *A treasure of Masdevallia: a monograph of the genus Masdevallia*. Volumes 21-26, Missouri Botanical Garden, USA.

Luer, C.A. (2000). *Icones Pleurothallidinarum XIX. Systematics of Masdevallia: Part one*. Missouri Botanical Garden, USA.

Luer, C.A. (2000). *Icones Pleurothallidinarum XXI. Systematics of Masdevallia: Part Two*. Missouri Botanical Garden, USA.

Luer, C.A. (2001). *Icones Pleurothallidinarum XXII. Systematics of Masdevallia: Part Three*. Missouri Botanical Garden, USA.

Préambule

Luer, C.A. (2002). *Icones Pleurothallidinarum XXIII. Systematics of Masdevallia: Part Four*. Missouri Botanical Garden, USA.

Luer, C.A. (2003). *Icones Pleurothallidinarum XXV. Systematics of Masdevallia: Part Five*. Missouri Botanical Garden, USA.

Ormerod, P. (1997). A Review of Coelogyne sect. Proliferae. *Austr. Orch. Rev.* 19-23.

Pearce, N.R. & Cribb, P.J. (2002). *The Orchids of Bhutan*. Royal Botanic Garden, Edinburgh. UK.

Seidenfaden, G. (1977/78). Orchid genera in Thailand: I-III. *Dansk Botanisk Arkiv* 29: 3-4.

Seidenfaden, G. (1988). Orchid genera in Thailand: XIV. Fifty-nine vandoid genera. *Opera Botanicam,* Vol 95.

Seidenfaden, G. (1992). The Orchids of Indochina. *Opera Botanica,* Vol 114.

Seidenfaden, G. & Wood, J.J. (1992). *The Orchids of Peninsular Malaysia and Singapore; a revision of R. E. Holttum: Orchids of Malaya*, Olsen & Olsen, Fredensborg, Denmark.

Stern, W.T. (1992). *Botanical Latin (fourth edition)*. David & Charles Publishers, Newton Abbot, Devon. UK.

The International Plant Names Index (2004). Published on the internet, http://www.ipni.org.

World Checklist of Monocots (2004). The Board of Trustees of the Royal Botanic Gardens, Kew. Published on the internet, www.kew.org/monocotChecklist/.

LISTA CITES - ORCHIDACEAE

PREÁMBULO

1. Antecedentes

En 1992 la Conferencia de las Partes en la Convención sobre el Comercio Internacional de Especies Amenazadas de Fauna y Flora Silvestres (CITES) aprobó la Resolución Conf. 8.19, en la que se solicita que se prepare una referencia normalizada sobre los nombres de las Orchidaceae.

Se encargó al Vicepresidente del Comité de Nomenclatura de la CITES que coordinase las tareas necesarias para preparar dicha referencia.

Se acordó abordar en primer lugar los géneros de orquídeas identificados como prioritarios en el Examen del Comercio Significativo de especies incluidas en el Apéndice II de CITES (CITES Doc. 8.31). Las listas (o partes de las mismas) se presentarían a la aprobación de la Conferencia de las Partes a medida que se fuesen preparando.

En su tercera reunión (Chiang Mai, Tailandia, noviembre de 1992), el Comité de Flora examinó detenidamente una propuesta del Vicepresidente del Comité de Nomenclatura sobre los posibles mecanismos para preparar una referencia normalizada. El Comité de Flora ratificó un procedimiento mediante el cual se efectuarían recopilaciones a partir de las publicaciones disponibles sobre el tema, que se introducirían en una base central de datos, y se presentaría a un Grupo Internacional de Expertos para efectuar consultas y tomar decisiones sobre los nombres que deberían utilizarse para los taxa en cuestión.

Durante la preparación de las listas se desplegaron esfuerzos para contactar a expertos nacionales de los Estados del área de distribución, utilizando como coordinadores a los representantes regionales ante el Comité de Flora. Este proceso tuvo poca acogida, y se espera que la publicación de las listas inspire una mayor participación en volúmenes venideros.

2. Programa informático

Soporte Físico: La base de datos se instaló en un Pentium III equipada con un software ALICE.

Sistema de base de datos: Se utilizó el sistema de base de datos ALICE para la recopilación de los datos. ALICE maneja datos sobre distribución, utilización, nombres comunes, descripciones, hábitats, sinónimos, y bibliografía, entre otros, de especies, subespecies o variedades. Permite al usuario confeccionar sus propios informes, por ejemplo, para listas de referencia estudios, monografías o listas de conservación.

El soporte lógico ALICE puede contactarse por correo electrónico a la dirección info@alicesoftware.com. Página en la Web: http://www.alicesoftware.com.

3. Procedimiento para la recopilación

- Las referencias preliminares fueron identificadas por especialistas en orquídeas del Real Jardín Botánico de Kew.
- Se estableció un Grupo Internacional de Expertos sobre orquídeas para que revisara cada una de las fases de la lista.
- Se introdujo la información en la base de datos taxonómica de ALICE y se preparó un informe preliminar.

Preámbulo

- Los informes preliminares respecto de cada género se distribuyeron al Grupo de expertos sobre orquídeas para que formulase comentarios sobre cualquier adición o enmienda.
- Las adiciones y enmiendas remitidas por el Grupo de expertos se introdujeron en la base de datos, vinculándolas a una referencia contenida en la bibliografía al final del informe sobre cada género. (No incluida en la presente lista, pero se guardan copias para referencia en el Real Jardín Botánico de Kew).
- Esta secuencia se repitió cinco veces para cada género, a fin de realizar consultas pormenorizadas con el Grupo.
- Se preparó un "borrador final" en el que se incluían todos los géneros y se distribuyó a otros expertos para que formulasen comentarios.
- Andre Schuiteman, Ed de Vogel y Johan Hermans participaron como editores externos para revisar la versión final.
- Las nuevas adiciones o enmiendas se incluyeron en la base de datos.
- El formato para la publicación se acordó con la Secretaría CITES, los informes se efectuaron en AWRITE y el material preparado para la cámara se elaboró utilizando Microsoft Word for Windows 2000.

4. Conservación

Durante el proceso de consultas con el Grupo de expertos sobre orquídeas, se solicitó también información sobre el estado de conservación de las especies en cuestión. Se transmitió al Grupo copia de las categorías de amenaza de la UICN actuales y propuestas.

5. Cómo emplear esta Lista

La idea es que esta Lista se utilice como referencia rápida para comprobar los nombres aceptados, los sinónimos y la distribución. Así, pues, la referencia se divide en tres partes principales:

Parte I: Todos los nombres de uso actual

Una lista de todos los nombres y sinónimos aceptados, en orden alfabético - un total de 1738 nombres (771 aceptados y 967 sinónimos).

Parte II: Nombres aceptados de uso actual

Listas independientes para cada género. En cada lista se presentan por orden alfabético los nombres aceptados, con información sobre sinónimos y distribución.

Parte III: Lista por países

Los nombres aceptados para todos los géneros incluidos en esta Lista se presentan por orden alfabético según el país de distribución.

6. Sistema de presentación utilizado en las Partes I, II y III

a) Los nombres aceptados se presentan en tipo de letra negrita y romano. Los sinónimos se presentan en letra cursiva.

b) Nombres duplicados

En la Parte I, el nombre del autor aparece después de cada taxón, cuando dicho taxón se cita en más de una ocasión p.e. *Aerides angustifolia* Hook.f., *Aerides angustifolia* (Blume) Lindl. non Hook.f. (a menos que el nombre del autor sea el mismo).

i) Donde un sinónimo aparece dos veces, pero se refiere a diferentes nombres aceptados, a saber, *Aerides latifolium*, (un sinónimo de ambas **Aerides odorata** y **Phalaenopsis deliciosa**), el nombre acompañado de un asterisco se refiere a la especie es más probable encontrar en el comercio, cuándo se sepa. Por ejemplo:

Todos los nombres	Nombre aceptado
Aerides latifolium (Thunb. ex Sw.) Sw.... ....**Aerides odorata***	
Aerides latifolium Thwaites non (Thunb. ex Sw.) Sw ....**Phalaenopsis deliciosa**	

*La especie que con mayor probabilidad se encontrará en el comercio (en este ejemplo, **Aerides odorata**).

ii) Si un nombre aceptado es igual al sinónimo, pero se refiere a especies diferentes, a saber, **Masdevallia meleagris** (nombre aceptado) y *Masdevallia meleagris* (un sinónimo de **Masdevallia picturata**), el nombre acompañado de un asterisco se refiere a la especie que es más probable encontrar en el comercio, cuando se sepa. Por ejemplo:

Todos los nombres	Nombre aceptado
Masdevallia meleagris Lindl.*	
Masdevallia meleagris Lindl. sensu Rchb.f. non Lindl. ...**Masdevallia picturata**	

*La especie que es más probable encontrar en el comercio (en este ejemplo, **Masdevallia meleagris**).

NB: En los ejemplos b)i) y b)ii) es preciso efectuar una doble verificación en lo que concierne a la distribución, como se indica en la Parte II. Por ejemplo, en el caso b)ii), si se dio el nombre '*Masdevallia meleagris*' y se sabe que la planta en cuestión procede de Guyana, querrá decir que la especie era **Masdevallia picturata** comercializada bajo el sinónimo *Masdevallia meleagris*. ya que **Masdevallia meleagris** se encuentra únicamente en Colombia.

c) En la lista se han incluido los híbridos naturales, indicados con el signo de multiplicar "×". Se presentan en las listas por orden alfabético.

d) Se han publicado muchos epítetos de especies de *Aerides* con terminaciones erróneas en latín. Éstas deben llevar la correspondiente forma femenina en latín, pues así es cómo se publicó el nombre genérico originalmente. No obstante, la mayor parte de las terminaciones de los epítetos de los nombres de especies se ha publicado en la forma neutra. Estos epítetos se han cambiado, para terminar con la forma femenina correcta, de conformidad con el Art. 32.5 del Código Internacional de Nomenclatura Botánica (2000).

La forma neutra sigue siendo de uso común, empleándose en los datos comerciales de CITES y en muchas obras de referencia importantes. Por eso, hemos incluido los epítetos con la forma neutra como sinónimos.

e) El CD-ROM contiene los siguientes archivos:

"CITESOrchidChecklist4.pdf", en formato de Adobe Acrobat®. Este archivo contiene el libro de la *CITES Orchid Checklist 4* y se puede visualizar con el programa Acrobat Reader®. Vd. deberá tener Adobe Acrobat Reader® instalado en su ordenador para ver este archivo (puede descargarse de www.adobe.com).

Se navega fácilmente por el archivo de Adobe Acrobat® con el uso de los marcadores dinámicos que se encuentran al margen izquierdo del documento, o con los hipervínculos incluidos dentro del mismo. Todos los hipervínculos dentro del documento están resaltados en azul.

"OrchidIntro.pdf", en formato de Adobe Acrobat®. Este archivo contiene el texto de introducción a los volúmenes 1-3 de las *CITES Orchid Checklists*. Se puede visualizar el archivo con el programa Acrobat Reader®. Vd. deberá tener Adobe Acrobat Reader® instalado en su ordenador para ver este archivo (puede descargarse de www.adobe.com).

"Vols I II and III combined_1.pdf", en formato de Adobe Acrobat®. Este archivo contiene una lista alfabética de todos los nombres y sinónimos aceptados para los géneros, igual que en los volúmenes 1-3 de las *CITES Orchid Checklists*. Se puede visualizar el archivo con el programa Acrobat Reader®. Vd. deberá tener Adobe Acrobat Reader® instalado en su ordenador para ver este archivo (puede descargarse de www.adobe.com).

"Vols I II and III combined_2.pdf", en formato de Adobe Acrobat®. Este archivo contiene una lista, en orden alfabético, de los nombres aceptados, sus sinónimos de uso actual, y la distribución de los géneros, igual que en los volúmenes 1-3 de las *CITES Orchid Checklists*. Se puede visualizar el archivo con el programa Acrobat Reader®. Vd. deberá tener Adobe Acrobat Reader® instalado en su ordenador para ver este archivo (puede descargarse de www.adobe.com).

"Vols I II and III combined_3.pdf", en formato de Adobe Acrobat®. Este archivo contiene los nombres aceptados de todos los géneros incluidos en los volúmenes 1-3 de las *CITES Orchid Checklists*, en orden alfabético por países de distribución. Se puede visualizar el archivo con el programa Acrobat Reader®. Vd. deberá tener Adobe Acrobat Reader® instalado en su ordenador para ver este archivo (puede descargarse de www.adobe.com).

7. Número de nombres incluidos para cada género:
Aerides (Aceptados: 23, Sinónimos: 290); *Coelogyne* (Aceptados: 194, Sinónimos: 287); *Comparettia* (Aceptados: 6, Sinónimos: 7); *Masdevallia* (Aceptados: 548, Sinónimos: 383).

8. Áreas geográficas
Para los nombres de los países se ha seguido la referencia oficial de las Naciones Unidas. *Terminology Bulletin*, United Nations, No.347, agosto de 1997, 1 - 41.

9. Orchidaceae controladas por la CITES
La familia Orchidaceae está incluida en el Apéndice II de CITES. Además, en el momento de esta publicación, están incluidos en el Apéndice I los siguientes taxa:

Aerangis ellisii
Dendrobium cruentum
Laelia jongheana
Laelia lobata
Paphiopedilum spp.
Peristeria elata
Phragmipedium spp.
Renanthera imschootiana

10. Abreviaturas, términos botánicos y expresiones latinas*

Estas abreviaturas, términos botánicos y expresiones latinas se han incluido como referencia útil, aunque no todas aparecen en la Lista. Nota: las palabras en *bastardilla* son latinas.

ambiguous name (nombre ambiguo) un nombre utilizado por distintos autores para diferentes taxa, convirtiéndose en una fuente de ambigüedad

anon. anonymous (anónimo) sin autor o autor desconocido

auct. *auctorum* de autores

CITES Convención sobre el Comercio Internacional de Especies Amenazadas de Fauna y Flora Silvestres

cultivar un ejemplar, o una agrupación de plantas, que tiene los mismos rasgos distintivos, y que se ha producido o se mantiene (se reproduce) en condiciones de cultivo

cultivation (cultivo) el cultivo de plantas mediante horticultura o jardinería; no extraída directamente del medio silvestre

descr. *descriptio*: la descripción de una especie o de otra unidad taxonómica

distribution (distribución) donde se encuentran las plantas (geográfica)

ed. editor

edn. edición (libro o revista)

eds. editores

epithet (epíteto) la última palabra de una especie, subespecie o variedad (etc.), por ejemplo: *speciosa* es el epíteto de la especie *Coelogyne speciosa*

escape (asilvestrada) una planta que ha traspasado los límites de su lugar de cultivo (p.e.: un jardín) y prospera entre la vegetación natural

ex *ex*: después; puede utilizarse entre los nombres de dos autores, el segundo de los cuales publicó el nombre indicado o sugerido por el primero

excl. *exclusus*: excluida

forma *forma*: una unidad taxonómica inferior al nivel de variedad

hort. *hortorum*: de jardines (horticultura); se cultiva o se encuentra en jardines; no se trata de una planta silvestre

ICNB (CINB) Código Internacional de Nomenclatura Botánica

in prep. en preparación

in sched. *in scheda*: en un espécimen de herbario o etiqueta

in syn. *in synonymia*: en sinonimia

incl. inclusive

ined. *ineditus*: inédito

introduction (introducción) una planta que se da en un país, o en cualquier otra localidad, por influencia antropogénica (intencionadamente o por casualidad); cualquier planta alóctona; que no sea nativa

key (clave) un sistema escrito utilizado para la identificación de organismos (p.e.: plantas)

leg. *legit*: él recolectó; el recolector o coleccionista

misspelling (error de ortografía) un nombre que se ha escrito incorrectamente; no se trata de un nombre nuevo o diferente

morphology (morfología) la forma y estructura de un organismo (p.e.: una planta)

name causing confusion (nombre que provoca confusión) un nombre que no se usa por no poder asignarse sin ambigüedad a un determinado taxón (p.e.: una especie vegetal)

native (nativo) un organismo (p.e.: una planta) que se da naturalmente en un país o región, etc.

naturalized (naturalizada) una planta alóctona (véase "introduction") o asilvestrada (véase "escape") pero que parece silvestre, y es capaz de reproducirse en su nuevo medio

nec *neque*: y no, tampoco

nom. *nomen* nombre

nom. ambig. *nomen ambiguum*: nombre ambiguo

nom. cons. prop. *nomen conservandum propositum*: nombre propuesto para la conservación con arreglo a lo dispuesto en el Código Internacional de Nomenclatura Botánica (ICBN)

nom. illeg. *nomen illegitimum*: nombre ilegítimo

nom. nud. *nomen nudum*: nombre publicado sin descripción

nomenclature (nomenclatura) rama de la ciencia que se ocupa de atribuir nombres a organismos (p.e.: plantas)

non *non*: no

only known from cultivation (sólo se conoce en cultivo) una planta que no se da en la naturaleza, únicamente en condiciones de cultivo

orthographic variant (variante ortográfica) una alternativa ortográfica del mismo nombre

p. p. *pro parte*: parcialmente, en parte

provisional name (nombre provisional) nombre asignado temporalmente hasta que se disponga de una descripción válida

sens. *sensu*: en el sentido de; la forma en que un autor interpreta o utiliza un nombre

sens. lat. *sensu lato*: en sentido amplio, un taxón (normalmente una especie) y todos sus taxa subordinados (p.e.: subspecies) y/u otros taxa a veces distintos

sic *sic*, utilizado después de una palabra que parece errónea o absurda, para dar a entender que se ha citado textualmente

spp. especies

ssp. subespecies

synonym (sinónimo) un nombre que se aplica a un taxón pero que no puede utilizarse por no ser el nombre aceptado – el sinónimo o los sinónimos forman la sinonimia

taxa plural de taxón

taxon (taxón) una determinada unidad de clasificación, p.e.: género, especie, subespecie

var. variedad

*Expresamos nuestro agradecimiento al Dr. Aaron Davis, Real Jardín Botánico de Kew, por proporcionar esta guía.

11. Bibliografía
Principales fuentes de referencia utilizadas para la recopilación de las listas:

Averyanov, L.V. & Averyanova, A.L. (2003). *Updated checklist of the orchids of Vietnam*. Vietnam National University Publishing House. Hanoi.

Bock, I. (1986). Revision der Gattung *Comparettia* Poepp. & Endl. (Teil 1*) Die Orchidee.* 37 (4): 192-196.

Bock, I. (1986). Revision der Gattung *Comparettia* Poepp. & Endl. (Teil 2). *Die Orchidee.* 37(5): 199-196.

Bock, I. (1986). Revision der Gattung *Comparettia* Poepp. & Endl. (Teil 3). *Die Orchidee.* 37(6): 255-263.

Brummit, R.K. & Powell C.E. (1992). *Authors of plant names.* Royal Botanic Gardens, Kew. UK.

Chen, S. [*et al.*]: (1999). *Angiospermae Flora reipublicae popularis Sinicae tomus 18. Monocotyledoneae.* Orchidaceae 2. Science Press, China.

Christenson, E.A. (1987). The taxonomy of Aerides & related genera in: Kamezo, S. & Tanaka, R. (eds), (1987). *Proceedings of the 12th World Orchid Conference*, World Orchid Conference, Tokyo, 12: 35-40.

Clayton, D. (2002). *The genus Coelogyne: a synopsis.* Royal Botanic Gardens, Kew. UK.

Comber, J.B. (2001). *Orchids of Sumatra.* Royal Botanic Gardens, Kew. UK.

Gagnepain, F. & Guillaumin, A. (1932 - 1934). *Orchidacees.* In Gagnepain, F (ed) (1908 – 1942). Flore Generale de L'Indo-Chine. 16: 142-647.

Cribb, P.J. & Wood, J.J. (1994). *A Checklist of the Orchids of Borneo.* Royal Botanic Gardens, Kew, UK.

Kraenzlin, F.W.L. (1925). *Repertorium Speciecum nobarum regni vegetabilis, Monographie der Guttungen Masdevallia.* Dahlem bei Berlin. Germany.

Luer, C.A. (1983 - 1998). *Thesaurus Masdevalliarum: a monograph of the genus Masdevallia.* Volumes 1-20(a), Missouri Botanical Garden, USA.

Luer, C.A. (1998 - 2001). *A treasure of Masdevallia: a monograph of the genus Masdevallia.* Volumes 21-26, Missouri Botanical Garden, USA.

Luer, C.A. (2000). *Icones Pleurothallidinarum XIX. Systematics of Masdevallia: Part one.* Missouri Botanical Garden, USA.

Luer, C.A. (2000). *Icones Pleurothallidinarum XXI. Systematics of Masdevallia: Part Two.* Missouri Botanical Garden, USA.

Luer, C.A. (2001). *Icones Pleurothallidinarum XXII. Systematics of Masdevallia: Part Three.* Missouri Botanical Garden, USA.

Luer, C.A. (2002). *Icones Pleurothallidinarum XXIII. Systematics of Masdevallia: Part Four.* Missouri Botanical Garden, USA.

Preámbulo

Luer, C.A. (2003). *Icones Pleurothallidinarum XXV. Systematics of Masdevallia: Part Five.* Missouri Botanical Garden, USA.

Ormerod, P. (1997). A Review of Coelogyne sect. Proliferae. *Austr. Orch. Rev.* 19-23.

Pearce, N.R. & Cribb, P.J. (2002). *The Orchids of Bhutan.* Royal Botanic Garden, Edinburgh. UK.

Seidenfaden, G. (1977/78). Orchid genera in Thailand: I-III. *Dansk Botanisk Arkiv* 29: 3-4.

Seidenfaden, G. (1988). Orchid genera in Thailand: XIV. Fifty-nine vandoid genera. *Opera Botanicam,* Vol 95.

Seidenfaden, G. (1992). The Orchids of Indochina. *Opera Botanica,* Vol 114.

Seidenfaden, G. & Wood, J.J. (1992). *The Orchids of Peninsular Malaysia and Singapore; a revision of R. E. Holttum: Orchids of Malaya,* Olsen & Olsen, Fredensborg, Denmark.

Stern, W.T. (1992). *Botanical Latin (fourth edition).* David & Charles Publishers, Newton Abbot, Devon. UK.

The International Plant Names Index (2004). Published on the internet, http://www.ipni.org.

World Checklist of Monocots (2004). The Board of Trustees of the Royal Botanic Gardens, Kew. Published on the internet, www.kew.org/monocotChecklist/.

PART I: ALL NAMES IN CURRENT USE
All names ordered alphabetically for the genera:

Aerides, *Coelogyne*, *Comparettia* and *Masdevallia*

PREMIÈRE PARTIE: TOUS LES NOMS D'USAGE COURANT
Par ordre alphabétique de tous les noms pour les genres:

Aerides, *Coelogyne*, *Comparettia* et *Masdevallia*

PARTE I: TODOS LOS NOMBRES DE USO ACTUAL
Todos los nombres en orden alfabético para los géneros:

Aerides, *Coelogyne*, *Comparettia* y *Masdevallia*

ALPHABETICAL LISTING OF ALL NAMES FOR THE GENERA:
Aerides, Coelogyne, Comparettia and *Masdevallia*

LISTES ALPHABETIQUES DE TOUS LES NOMS POUR LES GENRES:
Aerides, Coelogyne, Comparettia et *Masdevallia*

PRESENTACION POR ORDEN ALFABETICO DE ODOS LOS NOMBRES PARA EL GENERO:
Aerides, Coelogyne, Comparettia y *Masdevallia*

ALL NAMES TOUS LES NOMS TODOS LOS NOMBRES	ACCEPTED NAME NOM ACCEPTÉS NOMBRES ACEPTADOS
Aeeridium odorum	**Aerides odorata**
Aerides acuminatissima	**Thrixspermum acuminatissimum**
Aerides acuminatissimum	**Thrixspermum acuminatissimum**
Aerides affine	**Aerides multiflora**
Aerides affinis	**Aerides multiflora**
Aerides alba	**Aerides quinquevulnera**
Aerides album	**Aerides quinquevulnera**
Aerides amplexicaule	**Thrixspermum amplexicaule**
Aerides amplexicaulis	**Thrixspermum amplexicaule**
Aerides ampullacea	**Ascocentrum ampullaceum**
Aerides ampullaceum	**Ascocentrum ampullaceum**
Aerides anceps	**Thrixspermum anceps**
Aerides angustifolia (Blume) Lindl. non Hook.f.	**Thrixspermum angustifolium**
Aerides angustifolia Hook.f.	**Renanthera matutina***
Aerides angustifolium (Blume) Lindl. non Hook.f.	**Thrixspermum angustifolium**
Aerides angustifolium Hook.f.	**Renanthera matutina***
Aerides appendiculata	**Cleisostoma appendiculatum**
Aerides appendiculatum	**Cleisostoma appendiculatum**
Aerides arachnites (Blume) Lindl. non Sw.	**Thrixspermum centipeda**
Aerides arachnites Sw.	**Arachnis flos-aeris***
Aerides augustiana	
Aerides augustianum	**Aerides augustiana**
Aerides ballantiniana	**Aerides odorata**
Aerides ballantinianum	**Aerides odorata**
Aerides bernhardiana	**Aerides inflexa**
Aerides bernhardianum	**Aerides inflexa**
Aerides bicolor	**Vanda limbata**
Aerides biswasiana	**Papilionanthe biswasiana**
Aerides biswasianum	**Papilionanthe biswasiana**
Aerides borassii	**Cymbidium aloifolium**
Aerides brockessii	**Aerides crispa**
Aerides brookei	**Aerides crispa**
Aerides calceolare Buch.-Ham. ex Sm.	**Gastrochilus calceolaris***
Aerides calceolare Teijsm. & Binn. non Buch.-Ham. ex Sm.	**Ascochilus calceolaris**
Aerides calceolaris Buch.-Ham. ex Sm.	**Gastrochilus calceolaris***
Aerides calceolaris Teijsm. & Binn. non Buch.-Ham. ex Sm.	**Ascochilus calceolaris**
Aerides carnosa	**Phalaenopsis taenialis**
Aerides carnosum	**Phalaenopsis taenialis**
Aerides compressa	**Pteroceras compressum**
Aerides compressum	**Pteroceras compressum**
Aerides coriacea	**Angraecum coriaceum**
Aerides coriaceum	**Angraecum coriaceum**
Aerides cornuta	**Aerides odorata**

*For explanation see page 2, point 6
Voir les explications page 10, point 6
*Para mayor explicación, véase la página 20, point 6

ALL NAMES	ACCEPTED NAMES
Aerides cornutum	**Aerides odorata**
Aerides crassifolia	
Aerides crassifolium	**Aerides crassifolia**
Aerides crispa	
Aerides crispum	**Aerides crispa**
Aerides cristata	**Vanda cristata**
Aerides cristatum	**Vanda cristata**
Aerides cylindrica Hook.	**Papilionanthe vandarum***
Aerides cylindrica Lindl. ex Trimen non Hook.	**Papilionanthe subulata**
Aerides cylindricum Hook.	**Papilionanthe vandarum***
Aerides cylindricum Lindl. ex Trimen non Hook.	**Papilionanthe subulata**
Aerides dalzelliana	**Smithsonia viridiflora**
Aerides dalzellianum	**Smithsonia viridiflora**
Aerides dasypogon hort. ex Bateman non Sm.	**Stereochilus erinaceus**
Aerides dasypogon Sm.	**Gastrochilus dasypogon***
Aerides dayana	**Aerides odorata**
Aerides dayanum	**Aerides odorata**
Aerides decumbens	**Phalaenopsis parishii**
Aerides densiflora	**Robiquetia spatulata**
Aerides densiflorum	**Robiquetia spatulata**
Aerides difforme	**Ornithochilus difformis**
Aerides difformis	**Ornithochilus difformis**
Aerides diurna	**Pteroceras unguiculatum**
Aerides diurnum	**Pteroceras unguiculatum**
Aerides duquesnei	**Aerides odorata**
Aerides ellisii	**Aerides houlletiana**
Aerides elongata	**Renanthera elongata**
Aerides elongatum	**Renanthera elongata**
Aerides emarginata	**Ascochilus emarginatus**
Aerides emarginatum	**Ascochilus emarginatus**
Aerides emericii	
Aerides expansa	**Aerides crassifolia**
Aerides expansum	**Aerides crassifolia**
Aerides falcata	
Aerides falcatum	**Aerides falcata**
Aerides farmeri	**Aerides quinquevulnera**
Aerides fenzliana	**Aerides quinquevulnera**
Aerides fenzlianum	**Aerides quinquevulnera**
Aerides fieldingii	**Aerides rosea**
Aerides flabellata	
Aerides flabellatum	**Aerides flabellata**
Aerides flavescens	**Holcoglossum flavescens**
Aerides flavida	**Aerides odorata**
Aerides flavidum	**Aerides odorata**
Aerides flos-aeris	**Arachnis flos-aeris**
Aerides godefroyana	**Aerides multiflora**
Aerides godefroyanum	**Aerides multiflora**
Aerides greenii	**Papilionanthe greenii**
Aerides guttata	**Rhynchostylis retusa**
Aerides guttatum	**Rhynchostylis retusa**
Aerides houlletiana	
Aerides houlletianum	**Aerides houlletiana**
Aerides huttonii	**Aerides thibautiana**
Aerides hystrix (Blume) Lindl.	**Thrixspermum hystrix***
Aerides hystrix Lindl. non (Blume) Lindl.	**Ornithochilus difformis**
Aerides illustre	**Aerides maculosa**
Aerides illustris	**Aerides maculosa**
Aerides inflexa	

*For explanation see page 2, point 6
*Voir les explications page 10, point 6
*Para mayor explicación, véase la página 20, point 6

Part I: All Names / Tous les Noms / Todos los Nombres

ALL NAMES	ACCEPTED NAMES
Aerides inflexum	**Aerides inflexa**
Aerides × jansonii	
Aerides japonica	**Sedirea japonica**
Aerides japonicum	**Sedirea japonica**
Aerides jarckiana	**Aerides leeana**
Aerides jarckianum	**Aerides leeana**
Aerides jucunda	**Aerides odorata**
Aerides jucundum	**Aerides odorata**
Aerides krabiense	**Aerides krabiensis**
Aerides krabiensis	
Aerides larpentae	**Aerides falcata**
Aerides lasiopetala	**Eria lasiopetala**
Aerides lasiopetalum	**Eria lasiopetala**
Aerides latifolia (Thunb. ex Sw.) Sw.	**Aerides odorata***
Aerides latifolia Thwaites non (Thunb. ex Sw.) Sw.	**Phalaenopsis deliciosa**
Aerides latifolium (Thunb. ex Sw.) Sw.	**Aerides odorata***
Aerides latifolium Thwaites non (Thunb. ex Sw.) Sw.	**Phalaenopsis deliciosa**
Aerides lawrenceae	
Aerides lawrenciae	**Aerides lawrenceae**
Aerides leeana	
Aerides leeanum	**Aerides leeana**
Aerides leoniae	**insufficiently known**
Aerides leopardina	**Gastrochilus calceolaris**
Aerides leopardinum	**Gastrochilus calceolaris**
Aerides leopardorum	**Gastrochilus calceolaris**
Aerides lepida	**insufficiently known**
Aerides lepidum	**insufficiently known**
Aerides lindleyana	**Aerides crispa**
Aerides lindleyanum	**Aerides crispa**
Aerides lineare	**Aerides ringens**
Aerides linearis	**Aerides ringens**
Aerides lobbii Lem. non Teijsm. & Binn.	**Aerides multiflora***
Aerides lobbii Teijsm. & Binn.	**Thrixspermum calceolus**
Aerides longicornu	**Papilionanthe uniflora**
Aerides macrostachya	**Beclardia macrostachya**
Aerides macrostachyon	**Beclardia macrostachya**
Aerides maculata Buch.-Ham. ex Sm.	**Vanda spathulata**
Aerides maculata Llanos non Buch.-Ham. ex Sm.	**Aerides quinquevulnera***
Aerides maculatum Buch.-Ham. ex Sm.	**Vanda spathulata**
Aerides maculatum Llanos non Buch.-Ham. ex Sm.	**Aerides quinquevulnera***
Aerides maculosa	
Aerides maculosum	**Aerides maculosa**
Aerides margaritacea	**Aerides maculosa**
Aerides margaritaceum	**Aerides maculosa**
Aerides marginata	**Aerides quinquevulnera**
Aerides marginatum	**Aerides quinquevulnera**
Aerides matutina Blume non Willd.	**Renanthera matutina***
Aerides matutina Willd.	**Arachnis flos-aeris**
Aerides matutinum Blume non Willd.	**Renanthera matutina***
Aerides matutinum Willd.	**Arachnis flos-aeris**
Aerides mcmorlandii	
Aerides mendelii	**Aerides falcata**
Aerides micholitzii	**Aerides odorata**
Aerides minima	**Chroniochilus minimus**
Aerides minimum	**Chroniochilus minimus**
Aerides mitrata	**Seidenfadenia mitrata**
Aerides mitratum	**Seidenfadenia mitrata**

*For explanation see page 2, point 6
*Voir les explications page 10, point 6
*Para mayor explicación, véase la página 20, point 6

ALL NAMES	ACCEPTED NAMES
Aerides moschifera	**Arachnis flos-aeris**
Aerides multiflora Roxb.*	
Aerides multiflora auct. non Roxb.	**Aerides krabiensis**
Aerides multiflorum auct. non Roxb.	**Aerides krabiensis**
Aerides multiflorum Roxb.	**Aerides multiflora**
Aerides nobile	**Aerides odorata**
Aerides nobilis	**Aerides odorata**
Aerides obtusa	**Thrixspermum obtusum**
Aerides obtusum	**Thrixspermum obtusum**
Aerides odorata Lour.	
Aerides odorata Reinw. ex Blume non Lour.	**Aerides odorata**
Aerides odoratum Lour.	**Aerides odorata**
Aerides odoratum Reinw. ex Blume non Lour.	**Aerides odorata**
Aerides ortgiesiana	**Aerides quinquevulnera**
Aerides ortgiesianum	**Aerides quinquevulnera**
Aerides orthocentra	**Vanda testacea**
Aerides orthocentrum	**Vanda testacea**
Aerides pachyphylla	**Robiquetia pachyphylla**
Aerides pachyphyllum	**Robiquetia pachyphylla**
Aerides pallida (Blume) Lindl. non Roxb.	**Pteroceras pallidum**
Aerides pallida Blume non Roxb. nec (Blume) Lindl.	**Aerides timorana***
Aerides pallida Roxb.	**Micropera pallida**
Aerides pallidum (Blume) Lindl. non Roxb.	**Pteroceras pallidum**
Aerides pallidum Blume non Roxb. nec (Blume) Lindl. .	**Aerides timorana***
Aerides pallidum Roxb.	**Micropera pallida**
Aerides paniculata	**Cleisostoma paniculatum**
Aerides paniculatum	**Cleisostoma paniculatum**
Aerides pedunculata	**Papilionanthe pedunculata**
Aerides pedunculatum	**Papilionanthe pedunculata**
Aerides picotiana	**Aerides houlletiana**
Aerides picotianum	**Aerides houlletiana**
Aerides platychila	**Aerides houlletiana**
Aerides platychilum	**Aerides houlletiana**
Aerides praemorsa	**Rhynchostylis retusa**
Aerides praemorsum	**Rhynchostylis retusa**
Aerides purpurascens	**Thrixspermum purpurascens**
Aerides pusilla	**Grosourdya appendiculata**
Aerides pusillum	**Grosourdya appendiculata**
Aerides quinquevulnera	
Aerides quinquevulnerum	**Aerides quinquevulnera**
Aerides racemifera	**Cleisostoma racemiferum**
Aerides racemiferum	**Cleisostoma racemiferum**
Aerides radiata	**Bulbophyllum roxburghii**
Aerides radiatum	**Bulbophyllum roxburghii**
Aerides radicosa	**Aerides ringens**
Aerides radicosum	**Aerides ringens**
Aerides ramosa	**Staurochilus ramosus**
Aerides ramosum	**Staurochilus ramosus**
Aerides recurvipes	**Aerides leeana**
Aerides reichenbachii	**Aerides odorata**
Aerides retrofracta	**Aerides falcata**
Aerides retrofractum	**Aerides falcata**
Aerides retusa	**Rhynchostylis retusa**
Aerides retusum	**Rhynchostylis retusa**
Aerides reversa	**Aerides quinquevulnera**
Aerides reversum	**Aerides quinquevulnera**
Aerides rigida	**Acampe rigida**

*For explanation see page 2, point 6
*Voir les explications page 10, point 6
*Para mayor explicación, véase la página 20, point 6   33

ALL NAMES	ACCEPTED NAMES
Aerides rigidum	**Acampe rigida**
Aerides ringens	
Aerides roebelenii	**Aerides quinquevulnera**
Aerides rohaniana	**Aerides odorata**
Aerides rohanianum	**Aerides odorata**
Aerides rosea	
Aerides roseum	**Aerides rosea**
Aerides rostrata	**Micropera rostrata**
Aerides rostratum	**Micropera rostrata**
Aerides rubescens	
Aerides sanderiana	**Aerides lawrenceae**
Aerides sanderianum	**Aerides lawrenceae**
Aerides savageana	**Aerides quinquevulnera**
Aerides savageanum	**Aerides quinquevulnera**
Aerides schroederi	**Aerides maculosa**
Aerides shibatiana	**Aerides thibautiana**
Aerides shibatianum	**Aerides thibautiana**
Aerides siamense	**Aerides falcata**
Aerides siamensis	**Aerides falcata**
Aerides sillemiana	**Papilionanthe sillemiana**
Aerides sillemianum	**Papilionanthe sillemiana**
Aerides spicata	**Rhynchostylis retusa**
Aerides spicatum	**Rhynchostylis retusa**
Aerides spuria	**Dendrobium spurium**
Aerides spurium	**Dendrobium spurium**
Aerides suaveolens Blume non Roxb.	**Aerides odorata***
Aerides suaveolens Roxb.	**Pteroceras teres**
Aerides suavissima	**Aerides odorata**
Aerides suavissimum	**Aerides odorata**
Aerides subulata (Blume) Lindl.	**Thrixspermum subulatum***
Aerides subulata (J.Koenig) Schltr.	**Papilionanthe teres**
Aerides subulata (Sw.) Schltr. non (Blume) Lindl.	**Leucohyle subulata**
Aerides subulatum (Blume) Lindl.	**Thrixspermum subulatum***
Aerides subulatum (J.Koenig) Schltr.	**Papilionanthe subulata**
Aerides subulatum (Sw.) Schltr. non (Blume) Lindl.	**Leucohyle subulata**
Aerides sukauense	**Aerides sukauensis**
Aerides sukauensis	
Aerides sulingii	**Armodorum sulingii**
Aerides taeniale	**Phalaenopsis taenialis**
Aerides taenialis	**Phalaenopsis taenialis**
Aerides tenuifolia	**Cleisostoma tenuifolium**
Aerides tenuifolium	**Cleisostoma tenuifolium**
Aerides teres	**Pteroceras teres**
Aerides tessellata (Roxb.) Wight	**Vanda tessellata***
Aerides tessellata Thwaites non (Roxb.) Wight	**Vanda thwaitesii**
Aerides tessellatum (Roxb.) Wight	**Vanda tessellata*** *CITES Orchid*

Checklist 3 states that Aerides tessellatum was a synonym of Vanda spathulata. Aerides tessellatum is now considered a synonym of Vanda tessellata / Le Volume 3 de la Liste des Orchidées CITES mentionne que Aerides tessellatum est un synonyme de Vanda spathulata. Aerides tessellatum est actuellement considérée comme un synonyme de Vanda tessellata / En la CITES Orchid Checklist 3, Aerides tessellata se tenía por sinónimo de Vanda spathulata. En la actualidad, Aerides tessellata se considera sinónimo de Vanda tessellata.

Aerides tessellatum Thwaites non (Roxb.) Wight	**Vanda thwaitesii**
Aerides testacea	**Vanda testacea**
Aerides testaceum	**Vanda testacea**
Aerides teysmannii	**Macropodanthus teysmannii**
Aerides thibautiana	

*For explanation see page 2, point 6
*Voir les explications page 10, point 6
34   *Para mayor explicación, véase la página 20, point 6

ALL NAMES	ACCEPTED NAMES
Aerides thibautianum	**Aerides thibautiana**
Aerides thunbergii	**Neofinetia falcata**
Aerides timorana	
Aerides timoranum	**Aerides timorana**
Aerides trigona	**Aerides rosea**
Aerides trigonum	**Aerides rosea**
Aerides umbellata	**Gastrochilus acutifolius**
Aerides umbellatum	**Gastrochilus acutifolius**
Aerides uncinata	**Micropera uncinata**
Aerides uncinatum	**Micropera uncinata**
Aerides undulata	**Rhynchostylis retusa**
Aerides undulatum	**Rhynchostylis retusa**
Aerides uniflora	**Papilionanthe uniflora**
Aerides uniflorum	**Papilionanthe uniflora**
Aerides vandarum	**Papilionanthe vandarum**
Aerides veitchii	**Aerides multiflora**
Aerides virens	**Aerides odorata**
Aerides viridiflora	**Pteroceras viridiflorum**
Aerides viridiflorum	**Pteroceras viridiflorum**
Aerides warneri	**Aerides crispa**
Aerides wightiana auct. non Lindl.	**Vanda lilacina**
Aerides wightiana Lindl.	**Vanda testacea***
Aerides wightianum auct. non Lindl.	**Vanda lilacina**
Aerides wightianum Lindl.	**Vanda testacea***
Aerides williamsii	**Aerides rosea**
Aerides wilsoniana	**Aerides odorata**
Aerides wilsonianum	**Aerides odorata**
Aerides zollingeri	**Insufficiently known**
Angraecum nervosum	**Coelogyne rumphii**
Ascocentrum rubescens	**Aerides rubescens**
Broughtonia linearis	**Coelogyne ovalis**
Chelonanthera convallariifolia	**Coelogyne rochussenii**
Chelonanthera incrassata	**Coelogyne incrassata**
Chelonanthera longifolia	**Coelogyne longifolia**
Chelonanthera miniata	**Coelogyne miniata**
Chelonanthera speciosa	**Coelogyne speciosa**
Chelonistele vermicularis	**Coelogyne vermicularis**
Cleisostoma vacherotiana	**Aerides multiflora**
Coelogyne acutilabium	
Coelogyne advena	**Pholidota advena**
Coelogyne alata	**Coelogyne carinata**
Coelogyne alba	**Otochilus albus**
Coelogyne albobrunnea	
Coelogyne albolutea	
Coelogyne amplissima	**Chelonistele amplissima**
Coelogyne anceps	
Coelogyne angustifolia A.Rich.	**Coelogyne odoratissima**
Coelogyne angustifolia Ridl. non A.Rich. nec Wight	**Coelogyne trinervis***
Coelogyne angustifolia Wight non A.Rich.	**Coelogyne breviscapa**
Coelogyne annamensis (Lindl. & Rchb.f.) Rolfe	**Coelogyne assamica**
Coelogyne annamensis Ridl. non (Lindl. & Rchb.f.) Rolfe	**Coelogyne sanderae**
Coelogyne apiculata	**Panisea apiculata**
Coelogyne arthuriana	**Pleione maculata**
Coelogyne articulata	**Pholidota articulata**
Coelogyne arunachalensis	**Coelogyne fimbriata**
Coelogyne asperata	
Coelogyne assamica	
Coelogyne barbata	

*For explanation see page 2, point 6
*Voir les explications page 10, point 6
*Para mayor explicación, véase la página 20, point 6

Part I: All Names / Tous les Noms / Todos los Nombres

ALL NAMES	ACCEPTED NAMES
Coelogyne beccarii	
Coelogyne bella	**Coelogyne salmonicolor**
Coelogyne beyrodtiana	**Chelonistele sulphurea**
Coelogyne bicamerata	
Coelogyne biflora	**Panisea uniflora**
Coelogyne bihamata	**Coelogyne tenuis**
Coelogyne bilamellata	
Coelogyne bimaculata	**Coelogyne flexuosa**
Coelogyne birmanica	**Pleione praecox**
Coelogyne borneensis	
Coelogyne brachygyne	
Coelogyne brachyptera	
Coelogyne brevifolia	**Coelogyne punctulata**
Coelogyne brevilamellata	**Chelonistele brevilamellata**
Coelogyne breviscapa	
Coelogyne bruneiensis	
Coelogyne brunnea	**Coelogyne fuscescens**
Coelogyne buennemeyeri	
Coelogyne bulbocodioides	**Pleione bulbocodioides**
Coelogyne calcarata	
Coelogyne calceata	**Pholidota pallida**
Coelogyne calcicola	
Coelogyne caloglossa	
Coelogyne camelostalix	**Pholidota camelostalix**
Coelogyne candoonensis	
Coelogyne carinata	
Coelogyne carnea (Blume) Rchb.f.	**Pholidota carnea**
Coelogyne carnea Hook.f. non (Blume) Rchb.f.	**Coelogyne radicosa***
Coelogyne casta	**Coelogyne cumingii**
Coelogyne caulescens	**Bromheadia finlaysoniana**
Coelogyne celebensis	
Coelogyne chanii	
Coelogyne chinensis	**Pholidota chinensis**
Coelogyne chlorophaea	
Coelogyne chloroptera	
Coelogyne chrysotropis	
Coelogyne cinnamomea	**Coelogyne trinervis**
Coelogyne clarkei	**Coelogyne micrantha**
Coelogyne clemensii	
Coelogyne clypeata	**Pholidota gibbosa**
Coelogyne compressicaulis	
Coelogyne conchoidea	**Pholidota imbricata**
Coelogyne concinna	
Coelogyne conferta	**Coelogyne nitida**
Coelogyne confusa	
Coelogyne contractipetala	
Coelogyne convallariae	**Pholidota convallariae**
Coelogyne convallariifolia	**Coelogyne rochussenii**
Coelogyne corniculata	**Pholidota chinensis**
Coelogyne coronaria	**Eria coronaria**
Coelogyne corrugata	**Coelogyne nervosa**
Coelogyne corymbosa auct. non Lindl.	**Coelogyne nitida**
Coelogyne corymbosa Lindl.*	
Coelogyne crassifolia	**Chelonistele sulphurea**
Coelogyne crassiloba	
Coelogyne craticulaelabris	
Coelogyne cristata	
Coelogyne croockewitii	**Chelonistele sulphurea**

*For explanation see page 2, point 6
*Voir les explications page 10, point 6
36   *Para mayor explicación, véase la página 20, point 6

ALL NAMES	ACCEPTED NAMES
Coelogyne crotalina	**Pholidota imbricata**
Coelogyne cumingii	
Coelogyne cuneata	**Chelonistele sulphurea**
Coelogyne cuprea	
Coelogyne cycnoches	**Coelogyne fuscescens**
Coelogyne cymbidioides Rchb.f.	**Gynoglottis cymbidioides***
Coelogyne cymbidioides sensu Ridl. non Rchb.f.	**Coelogyne tomentosa**
Coelogyne dalatensis	**Coelogyne assamica**
Coelogyne darlacensis	**Coelogyne sanderae**
Coelogyne dayana	**Coelogyne pulverula**
Coelogyne decipiens	**Chelonistele sulphurea**
Coelogyne decora	**Coelogyne ovalis**
Coelogyne delavayi	**Pleione bulbocodioides**
Coelogyne densiflora	**Coelogyne tomentosa**
Coelogyne dichroantha	
Coelogyne diphylla	**Pleione maculata**
Coelogyne distans	
Coelogyne dulitensis	
Coelogyne eberhardtii	
Coelogyne ecarinata	
Coelogyne echinolabium	
Coelogyne edelfeldtii	**Coelogyne asperata**
Coelogyne elata Lindl.	**Coelogyne stricta***
Coelogyne elata sensu Hook. non Lindl.	**Coelogyne holochila**
Coelogyne elegans	**Cypripedium elegans**
Coelogyne elegantula	**Bletilla formosana**
Coelogyne elmeri	
Coelogyne endertii	
Coelogyne esquirolei	**Coelogyne esquirolii**
Coelogyne esquirolii	
Coelogyne exalata	
Coelogyne falcata	**Panisea uniflora**
Coelogyne filipeda	
Coelogyne fimbriata Lindl.*	
Coelogyne fimbriata auct. non Lindl.	**Coelogyne fuliginosa**
Coelogyne flaccida	
Coelogyne flavida Hook.f. ex Lindl.	**Coelogyne prolifera**
Coelogyne flavida sensu Seidenf. non Hook.f. ex Lindl. .	**Coelogyne schultesii***
Coelogyne fleuryi	**Coelogyne lawrenceana**
Coelogyne flexuosa	
Coelogyne foerstermannii	
Coelogyne fonstenebrarum	
Coelogyne formosa	
Coelogyne fragrans	
Coelogyne fuerstenbergiana	
Coelogyne fuliginosa	
Coelogyne fusca	**Otochilus fuscus**
Coelogyne fuscescens Lindl.*	
Coelogyne fuscescens Hook.f. non Lindl.	**Coelogyne assamica**
Coelogyne gardneriana	**Neogyna gardneriana**
Coelogyne genuflexa	
Coelogyne ghatakii	
Coelogyne gibbifera	
Coelogyne gibbosa	**Pholidota gibbosa**
Coelogyne glandulosa	
Coelogyne globosa	**Pholidota globosa**
Coelogyne gongshanensis	
Coelogyne goweri	**Coelogyne punctulata**

*For explanation see page 2, point 6
*Voir les explications page 10, point 6
*Para mayor explicación, véase la página 20, point 6

Part I: All Names / Tous les Noms / Todos los Nombres

ALL NAMES	ACCEPTED NAMES
Coelogyne graminifolia	**Coelogyne viscosa**
Coelogyne grandiflora	**Pleione grandiflora**
Coelogyne griffithii	
Coelogyne guamensis	
Coelogyne hajrae	
Coelogyne harana	
Coelogyne henryi	**Pleione bulbocodioides**
Coelogyne hirtella	
Coelogyne hitendrae	
Coelogyne holochila	
Coelogyne hookeriana	**Pleione hookeriana**
Coelogyne huettneriana Rchb.f.*	
Coelogyne huettneriana sensu Hook.f. non Rchb.f.	**Coelogyne flaccida**
Coelogyne humilis	**Pleione humilis**
Coelogyne imbricans	
Coelogyne imbricata	**Pholidota imbricata**
Coelogyne incrassata	
Coelogyne ingloria	**Chelonistele ingloria**
Coelogyne integerrima	
Coelogyne integra	
Coelogyne integrilabia	**Coelogyne fuscescens**
Coelogyne javanica	**Nervilia crociformis**
Coelogyne judithiae	
Coelogyne kaliana	
Coelogyne kelamensis	
Coelogyne kemiriensis	
Coelogyne khasiyana	**Pholidota articulata**
Coelogyne kinabaluensis	
Coelogyne kingii	**Coelogyne foerstermannii**
Coelogyne kutaiensis	**Chelonistele sulphurea**
Coelogyne lacinulosa	
Coelogyne lactea	**Coelogyne flaccida**
Coelogyne lagenaria	**Pleione × lagenaria**
Coelogyne lamellata	**Coelogyne macdonaldii**
Coelogyne lamellulifera	**Chelonistele lamellulifera**
Coelogyne laotica	**Coelogyne fimbriata**
Coelogyne latiloba	
Coelogyne lauterbachiana	**Coelogyne miniata**
Coelogyne lawrenceana	
Coelogyne lentiginosa	
Coelogyne leucantha	
Coelogyne leungiana	**Coelogyne fimbriata**
Coelogyne lockii	
Coelogyne loheri	
Coelogyne longeciliata	
Coelogyne longiana	**Coelogyne filipeda**
Coelogyne longibracteata	**Coelogyne cumingii**
Coelogyne longibulbosa	
Coelogyne longifolia	
Coelogyne longipes Lindl.*	
Coelogyne longipes sensu Hook.f. non Lindl.	**Coelogyne raizadae**
Coelogyne longirachis	
Coelogyne longpasiaensis	
Coelogyne loricata	**Pholidota imbricata**
Coelogyne lowii	**Coelogyne asperata**
Coelogyne lurida	**Chelonistele lurida**
Coelogyne lycastoides	
Coelogyne macdonaldii	

*For explanation see page 2, point 6
*Voir les explications page 10, point 6
*Para mayor explicación, véase la página 20, point 6

38

ALL NAMES	ACCEPTED NAMES
Coelogyne macrobulbon	**Coelogyne rochussenii**
Coelogyne macroloba	**Coelogyne gibbifera**
Coelogyne macrophylla	**Coelogyne asperata**
Coelogyne maculata	**Pleione maculata**
Coelogyne maingayi	**Coelogyne foerstermannii**
Coelogyne malintangensis	
Coelogyne malipoensis	
Coelogyne mandarinorum	**Ischnogyne mandarinorum**
Coelogyne marmorata	
Coelogyne marthae	
Coelogyne massangeana auct. non Rchb.f.	**Coelogyne kaliana**
Coelogyne massangeana Rchb.f.	**Coelogyne tomentosa***
Coelogyne mayeriana	
Coelogyne membranifolia	**Coelogyne septemcostata**
Coelogyne merrillii	
Coelogyne micholitziana	**Coelogyne beccarii**
Coelogyne micrantha	
Coelogyne miniata	
Coelogyne minutissima	**Dendrochilum sp. ?**
Coelogyne modesta	**Coelogyne prasina**
Coelogyne monilirachis	
Coelogyne monticola	
Coelogyne mooreana	
Coelogyne mossiae	
Coelogyne motleyi	
Coelogyne moultonii	
Coelogyne multiflora	
Coelogyne muluensis	
Coelogyne naja	
Coelogyne nervillosa	**Pholidota nervosa**
Coelogyne nervosa	
Coelogyne nigrofurfuracea	**Eriodes barbata**
Coelogyne nitida (Roxb.) Hook.f. non (Wall.ex D.Don) Lindl.	**Coelogyne punctulata**
Coelogyne nitida (Wall. ex D.Don) Lindl.*	
Coelogyne obtusifolia	
Coelogyne occultata	
Coelogyne ocellata	**Coelogyne punctulata**
Coelogyne ochracea	**Coelogyne nitida**
Coelogyne odoardii	
Coelogyne odoratissima	
Coelogyne oligantha	**Coelogyne carinata**
Coelogyne ovalis auct. non Lindl.	**Coelogyne pallens**
Coelogyne ovalis Lindl.*	
Coelogyne ovalis sensu Pfitzer & Kraenzl. non Lindl.	**Coelogyne fimbriata**
Coelogyne pachybulbon	**Coelogyne trinervis**
Coelogyne padangensis	
Coelogyne palaelabellata	**Calanthe aceras**
Coelogyne palawanensis	
Coelogyne palawensis	**Coelogyne guamensis**
Coelogyne pallens	
Coelogyne pallida	**Pholidota pallida**
Coelogyne pandurata	
Coelogyne papagena	**Coelogyne micrantha**
Coelogyne papillosa	
Coelogyne parishii	
Coelogyne parviflora	**Panisea demissa**
Coelogyne peltastes	
Coelogyne pendula	

*For explanation see page 2, point 6
*Voir les explications page 10, point 6
*Para mayor explicación, véase la página 20, point 6

Part I: All Names / Tous les Noms / Todos los Nombres

ALL NAMES	ACCEPTED NAMES
Coelogyne perakensis	Chelonistele sulphurea
Coelogyne phaiostele	Geesinkorchis phaiostele
Coelogyne pholas	Pholidota chinensis
Coelogyne pholidotoides	
Coelogyne picta	
Coelogyne pilosissima	Coelogyne ovalis
Coelogyne pinniloba	Chelonistele sulphurea
Coelogyne planiscapa	
Coelogyne plantaginea	Coelogyne rochussenii
Coelogyne platyphylla	Coelogyne celebensis
Coelogyne plicatissima	
Coelogyne pogonioides	Pleione bulbocodioides
Coelogyne porrecta	Otochilus porrecta
Coelogyne praecox	Pleione praecox
Coelogyne prasina	
Coelogyne primulina	Coelogyne fimbriata
Coelogyne prolifera Lindl.*	
Coelogyne prolifera sensu Gagnep. non Lindl.	Coelogyne filipeda
Coelogyne prolifera sensu Lindl.	Coelogyne schultesii
Coelogyne psectrantha	Coelogyne mooreana
Coelogyne psittacina	Coelogyne rumphii
Coelogyne pulchella	
Coelogyne pulverula sensu Lamb & C.L.Chan non Teijsm. & Binn.	Coelogyne rhabdobulbon
Coelogyne pulverula Teijsm. & Binn.*	
Coelogyne pumila	Dendrochilium pumilum
Coelogyne punctulata	
Coelogyne purpurascens	Adrorhizon purpurascens
Coelogyne pusilla.	Chelonistele sulphurea
Coelogyne pustulosa	Coelogyne asperata
Coelogyne quadrangularis	Coelogyne swaniana
Coelogyne quadratiloba	
Coelogyne quinquelamellata	
Coelogyne radicosa	
Coelogyne radicosus	Coelogyne radicosa
Coelogyne radioferens Ames & C.Schweinf.*	
Coelogyne radioferens sensu J.J.Sm. non Ames & C.Schweinf.	Coelogyne hirtella
Coelogyne radiosa	Coelogyne hirtella
Coelogyne raizadae	
Coelogyne ramosii	Chelonistele sulphurea
Coelogyne recurva	Pholidota recurva
Coelogyne reflexa	Coelogyne genuflexa
Coelogyne reichenbachiana	Pleione praecox
Coelogyne remediosae	Coelogyne remediosaie
Coelogyne remediosiae	
Coelogyne renae	
Coelogyne rhabdobulbon	
Coelogyne rhizomatosa	Coelogyne prasina
Coelogyne rhodeana	Coelogyne trinervis
Coelogyne rhombophora	Dendrochilum rhombophorum
Coelogyne richardsii	Chelonistele richardsii
Coelogyne ridleyana	Geesinkorchis phaiostele
Coelogyne ridleyi	Coelogyne sanderae
Coelogyne rigida	
Coelogyne rigidiformis	
Coelogyne rochussenii	
Coelogyne rossiana	Coelogyne trinervis
Coelogyne rubens	Calanthe rubens
Coelogyne rubra	Pholidota rubra

*For explanation see page 2, point 6
*Voir les explications page 10, point 6
*Para mayor explicación, véase la página 20, point 6

ALL NAMES	ACCEPTED NAMES
Coelogyne rumphii	
Coelogyne rupicola	
Coelogyne saigonensis	Coelogyne assamica
Coelogyne salmonicolor	
Coelogyne sanderae	
Coelogyne sanderiana	
Coelogyne sarawakensis	Chelonistele lurida
Coelogyne sarasinorum	Coelogyne carinata
Coelogyne schilleriana	
Coelogyne schultesii	
Coelogyne septemcostata	
Coelogyne siamensis	Coelogyne assamica
Coelogyne simplex	Coelogyne miniata
Coelogyne sparsa	
Coelogyne speciosa (Blume) Lindl.*	
Coelogyne speciosa sensu Ridl. non (Blume) Lindl.	Coelogyne septemcostata
Coelogyne squamulosa	
Coelogyne steenisii	
Coelogyne steffensii	Coelogyne rochussenii
Coelogyne stellaris	Coelogyne rochussenii
Coelogyne stenobulbon	
Coelogyne stenochila	
Coelogyne stenophylla	Coelogyne trinervis
Coelogyne stipitibulbum	Coelogyne radicosa
Coelogyne stricta	
Coelogyne suaveolens	
Coelogyne subintegra	Coelogyne exalata
Coelogyne sulphurea	Chelonistele sulphurea
Coelogyne sumatrana	Coelogyne testacea
Coelogyne susanae	
Coelogyne swaniana	
Coelogyne taronensis	
Coelogyne tenasserimensis	
Coelogyne tenompokensis	
Coelogyne tenuiflora	Pholidota gibbosa
Coelogyne tenuis	
Coelogyne testacea	
Coelogyne thailandica	Coelogyne quadratiloba
Coelogyne thuniana	Panisea uniflora
Coelogyne tiomanensis	
Coelogyne tomentosa	
Coelogyne tomiensis	Coelogyne tommii
Coelogyne tommii	
Coelogyne treutleri	Epigeneium treutleri
Coelogyne tricarinata	Coelogyne rigida
Coelogyne trifida	Coelogyne odoratissima
Coelogyne trilobulata	
Coelogyne trinervis	
Coelogyne triotos	Pholidota imbricata
Coelogyne triplicatula	
Coelogyne triptera	Epidendrum pygmaeum
Coelogyne trisaccata	Neogyna gardneriana
Coelogyne triuncialis	
Coelogyne truncicola	Coelogyne carinata
Coelogyne tumida	
Coelogyne undatialata	
Coelogyne undulata Rchb.f.	Pholidota rubra
Coelogyne undulata Wall. ex Pfitzer & Kraenzl. non Rchb.f. .	Coelogyne suaveolens*

*For explanation see page 2, point 6
*Voir les explications page 10, point 6
*Para mayor explicación, véase la página 20, point 6

Part I: All Names / Tous les Noms / Todos los Nombres

ALL NAMES	ACCEPTED NAMES
Coelogyne unguiculata	**Chelonistele unguiculata**
Coelogyne uniflora	**Panisea uniflora**
Coelogyne usitana	
Coelogyne ustulata	
Coelogyne vagans	**Coelogyne prasina**
Coelogyne vanoverberghii	
Coelogyne veitchii	
Coelogyne velutina	
Coelogyne ventricosa	**Pholidota ventricosa**
Coelogyne venusta	
Coelogyne vermicularis	
Coelogyne verrucosa	
Coelogyne virescens	
Coelogyne viscosa	
Coelogyne wallichiana	**Pleione praecox**
Coelogyne wallichii	**Pleione praecox**
Coelogyne wettsteiniana	**Coelogyne trinervis**
Coelogyne whitmeei	**Coelogyne lycastoides**
Coelogyne xanthoglossa	**Coelogyne xyrekes**
Coelogyne xerophyta	**Coelogyne fimbriata**
Coelogyne xylobioides	**Gynoglottis cymbidioides**
Coelogyne xyrekes	
Coelogyne yiii	
Coelogyne yunnanensis	**Pleione yunnanensis**
Coelogyne zahlbrucknerae	**Coelogyne marmorata**
Coelogyne zeylanica	
Coelogyne zhenkangensis	
Coelogyne zurowetzii	
Comparettia coccinea	
Comparettia cryptocera	**Comparettia falcata**
Comparettia erecta	**Comparettia falcata**
Comparettia falcata	
Comparettia ignea	
Comparettia macroplectron	
Comparettia × maloi	
Comparettia peruviana	**Comparettia coccinea**
Comparettia pulchella	**Comparettia falcata**
Comparettia rosea	**Comparettia falcata**
Comparettia speciosa	
Comparettia splendens	**Comparettia macroplectron**
Comparettia venezuelana	**Comparettia falcata**
Cymbidium evrardii	**Coelogyne assamica**
Cymbidium lineare	**Aerides ringens**
Cymbidium nitidum Roxb.	**Coelogyne punctulata**
Cymbidium nitidum Wall. ex D.Don non Roxb.	**Coelogyne nitida**
Cymbidium robustum	**Coelogyne asperata**
Cymbidium speciosissimum	**Coelogyne cristata**
Cymbidium stenopetalum	**Coelogyne longifolia**
Cymbidium strictum	**Coelogyne stricta**
Epidendrum aerides	**Aerides odorata**
Epidendrum geniculatum	**Aerides multiflora**
Epidendrum odoratum	**Aerides odorata**
Gastrochilus ringens	**Aerides ringens**
Gastrochilus speciosus	**Aerides maculosa**
Hologyne lauterbachiana	**Coelogyne miniata**
Hologyne miniata	**Coelogyne miniata**
Humboldtia plantaginea	**Masdevallia plantaginea**
Jostia teaguei	**Masdevallia teaguei**

*For explanation see page 2, point 6
*Voir les explications page 10, point 6
*Para mayor explicación, véase la página 20, point 6

Part I: All Names / Tous les Noms / Todos los Nombres

ALL NAMES	ACCEPTED NAMES
Limodorum latifolium	Aerides odorata
Lothiania bilabiata	Masdevallia plantaginea
Masdevallia abbreviata	
Masdevallia acaroi	
Masdevallia acrochordonia	
Masdevallia adamsii	
Masdevallia adrianae	
Masdevallia aenigma	
Masdevallia aequatorialis	Masdevallia pardina
Masdevallia aequiloba	Masdevallia civilis
Masdevallia affinis	Masdevallia laevis
Masdevallia agaster	
Masdevallia aguirrei	
Masdevallia akemiana	
Masdevallia albella	
Masdevallia albicans	Dryadella albicans
Masdevallia albida	Masdevallia infracta
Masdevallia alexandri	
Masdevallia alismifolia	
Masdevallia allenii	Trisetella triglochin
Masdevallia × alvaroi	
Masdevallia amabilis	
Masdevallia amaluzae	
Masdevallia amanda	
*Masdevallia amethystin*a	Porroglossum amethystinum
Masdevallia ametroglossa	
Masdevallia amoena	
Masdevallia amplexa	
Masdevallia ampullacea	
Masdevallia anachaeta	Diodonopsis anachaeta
Masdevallia anaristella.	Barbosella anaristella
Masdevallia anceps	
Masdevallia anchorifera	Scaphosepalum anchoriferum
Masdevallia andreettae	Dracula andreettae
Masdevallia andreettaeana	
Masdevallia andreettana	Masdevallia andreettaeana
Masdevallia anemone	
Masdevallia anfracta	
Masdevallia angulata	
Masdevallia angulifera	
Masdevallia anisomorpha	
Masdevallia anomala	
Masdevallia antioquiensis	Masdevallia molossus
Masdevallia antonii	
Masdevallia anura	Masdevallia molossoides
Masdevallia aops	Masdevallia klabochiorum
Masdevallia aperta	Pleurothallis tripterantha
Masdevallia aphanes	
Masdevallia apparitio	
Masdevallia approviata	Masdevallia coccinea
Masdevallia aptera	
Masdevallia arangoi	
Masdevallia argus	Zootrophion argus
Masdevallia ariasii	
Masdevallia aristata	Masdevallia infracta
Masdevallia armeniaca	Masdevallia coccinea
Masdevallia arminii	
Masdevallia aspera	Masdevallia paivaëana

*For explanation see page 2, point 6
*Voir les explications page 10, point 6
*Para mayor explicación, véase la página 20, point 6

ALL NAMES	ACCEPTED NAMES
Masdevallia asperrima	**Masdevallia melanoxantha**
Masdevallia assurgens	
Masdevallia asterotricha	
Masdevallia astuta	**Dracula astuta**
Masdevallia atahualpa	
Masdevallia atropurpurea	**Masdevallia bicolor**
Masdevallia atrosanguinea	**Masdevallia coccinea**
Masdevallia atroviolacea	**Masdevallia mooreana**
Masdevallia attenuata	
Masdevallia audax	
Masdevallia aurantiaca	**Masdevallia infracta**
Masdevallia aurea	
Masdevallia aureodactyla	**Masdevallia pachyura**
Masdevallia aurerosea	**Masdevallia bicolor**
Masdevallia auriculigera	**Dryadella auriculigera**
Masdevallia auropurpurea	**Masdevallia bicolor**
Masdevallia aurorae	
Masdevallia aviceps	**Dryadella aviceps**
Masdevallia ayabacana	
Masdevallia backhousiana	**Dracula chimaera**
Masdevallia bangii	
Masdevallia barlaeana	
Masdevallia barrowii	
Masdevallia bathyschista	**Masdevallia fasciata**
Masdevallia bella	**Dracula bella**
Masdevallia belua	
Masdevallia benedictii	**Dracula benedictii**
Masdevallia bennettii	
Masdevallia berthae	
Masdevallia bicolor	
Masdevallia bicornis	
Masdevallia biflora E.Morren	**Masdevallia bicolor***
Masdevallia biflora Regel non E.Morren	**Masdevallia caloptera**
Masdevallia bilabiata	**Masdevallia platyglossa**
Masdevallia blanda	**Masdevallia crassicaudis**
Masdevallia boddaertii	**Masdevallia ignea**
Masdevallia bogotensis	**Masdevallia coriacea**
Masdevallia boliviensis	
Masdevallia bomboiza	**Dracula lotax**
Masdevallia bonplandii	
Masdevallia borucana	**Masdevallia lata**
Masdevallia bottae	
Masdevallia bourdetteana	
Masdevallia braasii	**Masdevallia teaguei**
Masdevallia bradei	**Dryadella aviceps**
Masdevallia brachyantha	
Masdevallia brachyura	
Masdevallia brenneri	
Masdevallia brevis	**Scaphosepalum breve**
Masdevallia brockmuelleri	
Masdevallia bruckmuelleri	**Masdevallia coriacea**
Masdevallia bryophila	
Masdevallia buccinator	
Masdevallia bucculenta	
Masdevallia buchtienii	**Masdevallia scandens**
Masdevallia bulbophyllopsis	
Masdevallia burbidgeana	**Dracula erythrochaete**
Masdevallia burfordiensis	**Masdevallia angulata**

*For explanation see page 2, point 6
*Voir les explications page 10, point 6
*Para mayor explicación, véase la página 20, point 6

44

ALL NAMES	ACCEPTED NAMES
Masdevallia burianii	
Masdevallia burzlaffiana	**Masdevallia floribunda**
Masdevallia butcheri	**Trisetella tenuissima**
Masdevallia cacodes	
Masdevallia caesia	
Masdevallia calagrasalis	
Masdevallia callifera	**Dracula houtteana**
Masdevallia calocalix	
Masdevallia calocodon	**Masdevallia yungasensis**
Masdevallia caloptera	
Masdevallia calopterocarpa	**Masdevallia amanda**
Masdevallia calosiphon	
Masdevallia calura	
Masdevallia calyptrata	**Masdevallia corniculata**
Masdevallia campyloglossa Rchb.f.	
Masdevallia campyloglossa Rolfe non Rchb.f.	**Masdevallia campyloglossa**
Masdevallia candida Klotzsch & H.Karst. ex Rchb.f.	**Masdevallia tovarensis**
Masdevallia candida Linden non Klotzsch & H.Karst. ex Rchb.f. ...	**Masdevallia tovarensis**
Masdevallia capillaris	**Masdevallia plantaginea**
Masdevallia carderi	**Dracula inaequalis**
Masdevallia carderiopsis	**Dracula houtteana**
Masdevallia cardiantha	
Masdevallia carinata	**Dryadella zebrina**
Masdevallia carmenensis	
Masdevallia carnosa	
Masdevallia carolloi	**Masdevallia sernae**
Masdevallia carpishica	
Masdevallia carpophora	**Pleurothallis tripterantha**
Masdevallia carruthersiana	
Masdevallia casta	**Masdevallia tubulosa**
Masdevallia castor	
Masdevallia catapheres	
Masdevallia caudata Lindl.	
Masdevallia caudivolvula	
Masdevallia cayennensis	**Masdevallia cuprea**
Masdevallia cerastes	
Masdevallia chaetostoma	
Masdevallia chaparensis	
Masdevallia chasei	
Masdevallia chestertonii	**Dracula chestertonii**
Masdevallia chimaera Linden & André non Rchb.f.	**Dracula nycterina**
Masdevallia chimaera Rchb.f.	**Dracula chimaera***
Masdevallia chimboënsis	
Masdevallia chiquindensis	**Masdevallia leucantha**
Masdevallia chloracra	**Masdevallia striatella**
Masdevallia chlorotica	**Masdevallia laevis**
Masdevallia chontalensis	
Masdevallia chrysochaete	**Masdevallia strumifera**
Masdevallia chrysoneura	**Masdevallia uncifera**
Masdevallia chuspipatae	
Masdevallia cinnabarina	**Masdevallia coccinea**
Masdevallia cinnamomea	
Masdevallia citrinella	
Masdevallia civilis	
Masdevallia clandestina	
Masdevallia cleistogama	
Masdevallia cloesii	
Masdevallia cocapatae	

*For explanation see page 2, point 6
*Voir les explications page 10, point 6
*Para mayor explicación, véase la página 20, point 6

Part I: All Names / Tous les Noms / Todos los Nombres

ALL NAMES	ACCEPTED NAMES
Masdevallia coccinea Linden ex Lindl.*	
Masdevallia coccinea Regel non Linden ex Lindl.	**Masdevallia ignea**
Masdevallia coerulescens	**Masdevallia coccinea**
Masdevallia colibri	**Masdevallia trochilus**
Masdevallia collantesii	
Masdevallia collina	
Masdevallia colombiana	**Porroglossum mordax**
Masdevallia colossus	
Masdevallia concinna	
Masdevallia condorensis	
Masdevallia confusa	**Masdevallia laevis**
Masdevallia constricta	
Masdevallia copiosa	**Masdevallia hians**
Masdevallia corazonica	
Masdevallia cordeliana	
Masdevallia corderoana	
Masdevallia coriacea	
Masdevallia corniculata	
Masdevallia cosmia	
Masdevallia costaricensis	**Masdevallia marginella**
Masdevallia cranion	
Masdevallia crassicaudis	
Masdevallia crassicaulis	**Masdevallia crassicaudis**
Masdevallia crenulata	**Dryadella crenulata**
Masdevallia crescenticola	
Masdevallia cretata	
Masdevallia cryptocopis	**Masdevallia picturata**
Masdevallia cucullata	
Masdevallia cucutillensis	**Masdevallia caudata**
Masdevallia culex	**Pleurothallis macroblepharis**
Masdevallia cuprea	
Masdevallia cupularis	
Masdevallia curtipes	
Masdevallia cyathogastra	**Masdevallia nidifica**
Masdevallia cyclotega	
Masdevallia cylix	
Masdevallia dalessandroi	
Masdevallia dalstroemii	
Masdevallia datura	
Masdevallia davisii	
Masdevallia dayana	**Zootrophion dayanum**
Masdevallia deceptrix	
Masdevallia decumana	
Masdevallia deformis	
Masdevallia delhierroi	
Masdevallia delphina	
Masdevallia deltoidea	**Dracula deltoidea**
Masdevallia demissa	
Masdevallia deniseana	
Masdevallia denisonii	**Masdevallia coccinea**
Masdevallia densiflora	
Masdevallia deorsum	**Masdevallia caesia**
Masdevallia dermatantha	**Masdevallia campyloglossa**
Masdevallia descendens	
Masdevallia diantha	**Masdevallia chontalensis**
Masdevallia didyma	**Trisetella didyma**
Masdevallia dimorphotricha	
Masdevallia discoidea	

*For explanation see page 2, point 6
*Voir les explications page 10, point 6
*Para mayor explicación, véase la página 20, point 6

46

ALL NAMES	ACCEPTED NAMES
Masdevallia discolor	
Masdevallia dispar	**Masdevallia sanctae-fidei**
Masdevallia diversifolia	**Masdevallia parvula**
Masdevallia dodsonii	**Dracula dodsonii**
Masdevallia dolichopoda	**Dryadella dolichopoda**
Masdevallia don-quijote	
Masdevallia dorisiae	
Masdevallia draconis	
Masdevallia dreisei	
Masdevallia dressleri	**Trisetella dressleri**
Masdevallia dryada	
Masdevallia dudleyi	
Masdevallia dunstervillei	
Masdevallia dura	
Masdevallia dynastes	
Masdevallia eburnea	
Masdevallia ecaudata	**Masdevallia tubuliflora**
Masdevallia echidna	**Porroglossum echidna**
Masdevallia echinata	**Masdevallia rosea**
Masdevallia echinocarpa	**Diodonopsis erinacea**
Masdevallia echo	
Masdevallia eclyptrata	**Masdevallia corniculata**
Masdevallia eduardi	**Porroglossum eduardi**
Masdevallia edwallii	**Dryadella edwallii**
Masdevallia ejiriana	
Masdevallia elachys	
Masdevallia elata	**Dryadella elata**
Masdevallia elegans	
Masdevallia elephanticeps	
Masdevallia ellipes	**Masdevallia peristeria**
Masdevallia empusa	
Masdevallia enallax	**Masdevallia epallax**
Masdevallia encephala	
Masdevallia endotrachys	**Masdevallia bonplandii**
Masdevallia ensata	
Masdevallia epallax	
Masdevallia ephelota	
Masdevallia ephippium	**Masdevallia trochilus**
Masdevallia erinacea	**Diodonopsis erinacea**
Masdevallia erythrochaete	**Dracula erythrochaete**
Masdevallia espirito-santensis	**Dryadella espirito-santensis**
Masdevallia estradae	
Masdevallia eucharis	
Masdevallia eumeces	
Masdevallia eumeliae	
Masdevallia eurynogaster	
Masdevallia exaltata	**Masdevallia deformis**
Masdevallia excelsior	
Masdevallia exigua	**Diodonopsis pygmaea**
Masdevallia exilipes	**Masdevallia klabochiorum**
Masdevallia expansa	
Masdevallia expers	
Masdevallia exquisita	
Masdevallia falcago	
Masdevallia fasciata	
Masdevallia felix	**Dracula felix**
Masdevallia fertilis	**Masdevallia campyloglossa**
Masdevallia figueroae	

*For explanation see page 2, point 6
*Voir les explications page 10, point 6
*Para mayor explicación, véase la página 20, point 6

Part I: All Names / Tous les Noms / Todos los Nombres

ALL NAMES	ACCEPTED NAMES
Masdevallia filamentosa	**Masdevallia pumila**
Masdevallia filaria	
Masdevallia fimbriata	**Pleurothallis setosa**
Masdevallia fissa	**Masdevallia heteroptera**
Masdevallia flaccida	**Masdevallia uncifera**
Masdevallia flammula	**Masdevallia amabilis**
Masdevallia flaveola	
Masdevallia floribunda Lindl.	
Masdevallia floribunda Ames non Lindl.	**Masdevallia floribunda**
Masdevallia foeda	**Masdevallia menatoi**
Masdevallia foetens	
Masdevallia fonsecae	**Masdevallia attenuata**
Masdevallia forgetiana	**Masdevallia infracta**
Masdevallia formosa	
Masdevallia fosterae	
Masdevallia fractiflexa	
Masdevallia fragrans	
Masdevallia frilehmannii	
Masdevallia frontinoënsis	**Masdevallia herradurae**
Masdevallia fuchsii	
Masdevallia fuliginosa	**Dracula radiella**
Masdevallia fulvescens	
Masdevallia funebris	**Masdevallia reichenbachiana**
Masdevallia galeottiana	**Masdevallia floribunda**
Masdevallia garciae	
Masdevallia gargantua	
Masdevallia gaskelliana	**Dracula erythrochaete**
Masdevallia geminiflora	
Masdevallia gemmata	**Trisetella gemmata**
Masdevallia gerlachii	**Masdevallia macroglossa**
Masdevallia gibberosa	**Scaphosepalum gibberosum**
Masdevallia gigas	**Dracula gigas**
Masdevallia gilbertoi	
Masdevallia glandulosa	
Masdevallia glomerosa	
Masdevallia gloriae	
Masdevallia glossacles	**Masdevallia mentosa**
Masdevallia gnoma	
Masdevallia goliath	
Masdevallia gomes-ferreirae	**Dryadella gomes-ferreirae**
Masdevallia gomeziana	**Masdevallia laevis**
Masdevallia gorgo	**Dracula astuta**
Masdevallia gorgona	**Dracula gorgona**
Masdevallia gracilenta	**Zootrophion gracilentum**
Masdevallia gracilior	**Masdevallia lenae**
Masdevallia graminea	
Masdevallia grandiflora	**Masdevallia pumila**
Masdevallia grossa	**Masdevallia ophioglossa**
Masdevallia guatemalensis	**Dryadella guatemalensis**
Masdevallia guayanensis	
Masdevallia guerrieroi	
Masdevallia gustavii	**Masdevallia amanda**
Masdevallia gutierrezii	
Masdevallia guttulata Rchb.f.	
Masdevallia guttulata Rolfe non Rchb.f.	**Masdevallia guttulata**
Masdevallia haematocantha	**Masdevallia amanda**
Masdevallia haematosticta.	**Masdevallia peristeria**
Masdevallia hajekii	**Masdevallia chaparensis**

*For explanation see page 2, point 6
*Voir les explications page 10, point 6
*Para mayor explicación, véase la página 20, point 6

ALL NAMES	ACCEPTED NAMES
Masdevallia harlequina	
Masdevallia harryana	**Masdevallia coccinea**
Masdevallia hartmanii	
Masdevallia heideri	
Masdevallia helenae	
Masdevallia helgae	
Masdevallia henniae	
Masdevallia hepatica	**Masdevallia cuprea**
Masdevallia hercules	
Masdevallia herradurae	
Masdevallia herzogii	**Masdevallia bicolor**
Masdevallia heteromorpha	**Masdevallia heteroptera**
Masdevallia heteroptera	
Masdevallia heterotepala	**Masdevallia campyloglossa**
Masdevallia hians	
Masdevallia hieroglyphica	
Masdevallia hirtzii	
Masdevallia hoeijeri	**Diodonopsis hoeijeri**
Masdevallia hoppii	**Masdevallia pachyantha**
Masdevallia hornii	**Pleurothallis yupanki**
Masdevallia horrida	**Diodonopsis erinacea**
Masdevallia hortensis	
Masdevallia houtteana	**Dracula houtteana**
Masdevallia hubeinii	
Masdevallia huebneri	**Trisetella triglochin**
Masdevallia huebschiana	**Masdevallia polysticta**
Masdevallia humilis	**Masdevallia zahlbruckneri**
Masdevallia hydrae	
Masdevallia hylodes	
Masdevallia hymenantha	
Masdevallia hypodiscus	**Zootrophion hypodiscus**
Masdevallia hystrix	
Masdevallia icterina	
Masdevallia idae	
Masdevallia ignea	
Masdevallia immensa	
Masdevallia impostor	
Masdevallia inaequalis	**Dracula inaequalis**
Masdevallia indecora	
Masdevallia inflata	**Masdevallia corniculata**
Masdevallia infracta	
Masdevallia ingridiana	
Masdevallia instar	
Masdevallia invenusta	**Masdevallia bulbophyllopsis**
Masdevallia ionocharis	
Masdevallia irapana	
Masdevallia iricolor	**Dracula iricolor**
Masdevallia iris	
Masdevallia ishikoi	
Masdevallia isos	
Masdevallia jalapensis	**Pleurothallis jalapensis**
Masdevallia janetiae	**Dracula janetiae**
Masdevallia jarae	
Masdevallia jimenezii	**Masdevallia empusa**
Masdevallia johannis	**Dracula pusilla**
Masdevallia josei	
Masdevallia juan-albertoi	
Masdevallia jubar	**Masdevallia tridens**

*For explanation see page 2, point 6
*Voir les explications page 10, point 6
*Para mayor explicación, véase la página 20, point 6

ALL NAMES	ACCEPTED NAMES
Masdevallia kalbreyeri	**Masdevallia urceolaris**
Masdevallia karineae	
Masdevallia kautskyi	**Dryadella auriculigera**
Masdevallia klabochiorum	
Masdevallia kuhniorum	
Masdevallia kyphonantha	
Masdevallia lactea	**Dracula velutina**
Masdevallia laeta	**Masdevallia coccinea**
Masdevallia laevis	
Masdevallia lamia	
Masdevallia lamprotyria	
Masdevallia lankesterana	
Masdevallia lansbergii	
Masdevallia lappifera	
Masdevallia lata	
Masdevallia lateritia	**Masdevallia coccinea**
Masdevallia laucheana	
Masdevallia lawrencei	**Masdevallia guttulata**
Masdevallia leathersii	
Masdevallia lehmannii	
Masdevallia lenae	
Masdevallia leonardoi	
Masdevallia leonii	
Masdevallia leontoglossa	
Masdevallia lepida	**Masdevallia laevis**
Masdevallia leptoura	
Masdevallia leucantha	
Masdevallia leucophaea	**Masdevallia boliviensis**
Masdevallia lewisii	
Masdevallia × ligiae	
Masdevallia lilacina	
Masdevallia lilianae	
Masdevallia lilliputana	**Dryadella lilliputana**
Masdevallia lima	**Scaphosepalum lima**
Masdevallia limax	
Masdevallia linearifolia	**Dryadella linearifolia**
Masdevallia lindeniana	**Masdevallia floribunda**
Masdevallia lindenii	**Masdevallia coccinea**
Masdevallia lineolata	
Masdevallia lintricula	
Masdevallia livingstoneana	
Masdevallia longicaudata	**Masdevallia infracta**
Masdevallia longiflora Cogn.	**Masdevallia coccinea**
Masdevallia longiflora Kraenzl. non Cogn.	**Barbosella cucullata**
Masdevallia lotax	**Dracula lotax**
Masdevallia loui	
Masdevallia lowii	**Dracula platycrater**
Masdevallia lucernula	
Masdevallia ludibunda	
Masdevallia ludibundella	
Masdevallia × lueri	**Masdevallia × senghasiana**
Masdevallia luziae-mariae	
Masdevallia lychniphora	
Masdevallia lynniana	
Masdevallia macrochila	**Dracula chestertonii**
Masdevallia macrodactyla	**Scaphosepalum macrodactylum**
Masdevallia macrogenia	
Masdevallia macroglossa	

*For explanation see page 2, point 6
*Voir les explications page 10, point 6
*Para mayor explicación, véase la página 20, point 6

ALL NAMES	ACCEPTED NAMES
Masdevallia macropus	
Masdevallia macrura	
Masdevallia maculata	
Masdevallia maculigera	Masdevallia laevis
Masdevallia maduroi	
Masdevallia mallii	
Masdevallia maloi	
Masdevallia manarana	Masdevallia guayanensis
Masdevallia manchinazae	
Masdevallia mandarina	
Masdevallia manningii	Masdevallia cuprea
Masdevallia manoloi	
Masdevallia manta	
Masdevallia margaretae	Masdevallia carruthersiana
Masdevallia marginella	
Masdevallia marizae	
Masdevallia marthae	
Masdevallia martineae	
Masdevallia martiniana	
Masdevallia mascarata	
Masdevallia mastodon	
Masdevallia mataxa	
Masdevallia maxilimax	
Masdevallia maxillariiformis	Masdevallia strumifera
Masdevallia mayaycu	
Masdevallia medellinensis	Dracula radiosa
Masdevallia meiracyllium	Dryadella meiracyllium
Masdevallia medinae	
Masdevallia medusa	
Masdevallia megaloglossa	Masdevallia vargasii
Masdevallia mejiana	
Masdevallia melanoglossa	
Masdevallia melanopus	
Masdevallia melanoxantha	
Masdevallia meleagris Lindl.*	
Masdevallia meleagris Lindl. sensu Rchb.f. non Lindl ...	Masdevallia picturata
Masdevallia melina	Masdevallia lilianae
Masdevallia melloi	Dryadella melloi
Masdevallia menatoi	
Masdevallia mendozae	
Masdevallia mentosa	
Masdevallia merinoi	
Masdevallia metallica	Masdevallia caesia
Masdevallia mezae	
Masdevallia microglochin	Dracula velutina
Masdevallia microptera	
Masdevallia microsiphon	
Masdevallia midas	
Masdevallia mijahuangae	Masdevallia cyclotega
Masdevallia milagroi	
Masdevallia militaris	Masdevallia coccinea
Masdevallia miniata	Masdevallia coccinea
Masdevallia minuta	
Masdevallia misasii	
Masdevallia molossoides	
Masdevallia molossus	
Masdevallia × monicana	
Masdevallia monogona	

*For explanation see page 2, point 6
*Voir les explications page 10, point 6
*Para mayor explicación, véase la página 20, point 6

ALL NAMES	ACCEPTED NAMES
Masdevallia mooreana	
Masdevallia mopsus	**Dracula mopsus**
Masdevallia mordax	**Porroglossum mordax**
Masdevallia morenoi	**Masdevallia schizopetala**
Masdevallia morochoi	
Masdevallia mosquerae	**Dracula houtteana**
Masdevallia moyobambae	**Masdevallia weberbaueri**
Masdevallia murex	
Masdevallia muriculata	**Diodonopsis pygmaea**
Masdevallia musaica	**Masdevallia coccinea**
Masdevallia muscosa	**Porroglossum muscosum**
Masdevallia mutica	
Masdevallia myriostigma	**Masdevallia floribunda**
Masdevallia × mystica	
Masdevallia naranjapatae	
Masdevallia navicularis Garay & Dunst.*	
Masdevallia navicularis Kraenzl. non Garay & Dunst.nom.nud.	**Scaphosepalum anchoriferum**
Masdevallia nebulina	
Masdevallia neglecta	**Masdevallia rufescens**
Masdevallia newmaniana	
Masdevallia nicaraguae	
Masdevallia nidifica	
Masdevallia niesseniae	
Masdevallia nigricans	
Masdevallia nikoleana	
Masdevallia nitens	
Masdevallia nivalis	**Masdevallia hians**
Masdevallia nivea	
Masdevallia norae	
Masdevallia normannii	**Masdevallia reichenbachiana**
Masdevallia norops	
Masdevallia notosibirica	
Masdevallia nutans	**Diodonopsis anachaeta**
Masdevallia nycterina	**Dracula nycterina**
Masdevallia obrieniana	**Dryadella aviceps**
Masdevallia obscurans	
Masdevallia ocanensis	**Masdevallia picturata**
Masdevallia ochracea	**Masdevallia coriacea**
Masdevallia ochthodes	**Scaphosepalum verrucosum**
Masdevallia odontocera	
Masdevallia odontochila	**Masdevallia cupularis**
Masdevallia odontopetala	
Masdevallia oligantha	**Masdevallia amanda**
Masdevallia olivacea	**Masdevallia angulifera**
Masdevallia olmosii	**Masdevallia zahlbruckneri**
Masdevallia osmariniana	**Dryadella osmariniana**
Masdevallia omorenoi	
Masdevallia ophioglossa	
Masdevallia oreas	
Masdevallia ortalis	
Masdevallia ortgiesiana	**Masdevallia campyloglossa**
Masdevallia oscarii	
Masdevallia oscitans	
Masdevallia os-draconis	
Masdevallia ostaurina	
Masdevallia os-viperae	
Masdevallia ova-avis	
Masdevallia oxapampaensis	

*For explanation see page 2, point 6
*Voir les explications page 10, point 6
*Para mayor explicación, véase la página 20, point 6

52

ALL NAMES	ACCEPTED NAMES
Masdevallia pachyantha	
Masdevallia pachygyne	
Masdevallia pachysepala	
Masdevallia pachyura	
Masdevallia paisbambae	Masdevallia klabochiorum
Masdevallia paivaëana	
Masdevallia pallida	Masdevallia xanthina
Masdevallia palmensis	Masdevallia fasciata
Masdevallia panamensis	Masdevallia livingstoneana
Masdevallia pandurilabia	
Masdevallia panguiënsis	
Masdevallia pantex	Trisetella pantex
Masdevallia pantherina	Masdevallia laevis
Masdevallia pantomima	
Masdevallia papillosa	
Masdevallia paquishae	
Masdevallia pardina	
Masdevallia paranaensis	Dryadella paranaensis
Masdevallia × parlatoreana	Masdevallia × splendida
Masdevallia parvula	
Masdevallia pastensis	Masdevallia uncifera
Masdevallia pastinata	
Masdevallia patchicutzae	
Masdevallia patriciana	
Masdevallia patula	
Masdevallia paulensis	Dryadella aviceps
Masdevallia peristeria	
Masdevallia pernix	
Masdevallia perpusilla	Dryadella perpusilla
Masdevallia persicina	
Masdevallia peruviana	Masdevallia bicolor
Masdevallia pescadoënsis	
Masdevallia petiolaris	Masdevallia laevis
Masdevallia phacopsis	
Masdevallia phasmatodes	
Masdevallia phlogina	
Masdevallia phoenix	
Masdevallia picea	
Masdevallia picta	
Masdevallia picturata	
Masdevallia pileata	
Masdevallia pinocchio	
Masdevallia planadensis	
Masdevallia plantaginea	
Masdevallia platycrater	Dracula platycrater
Masdevallia platyglossa	
Masdevallia platyrachys	Pleurothallis endotrachys
Masdevallia pleurothalloides	
Masdevallia plynophora	
Masdevallia × polita	
Masdevallia pollux	
Masdevallia polyantha	Masdevallia schlimii
Masdevallia polychroma	
Masdevallia polyphemus	Dracula polyphemus
Masdevallia polysticta	
Masdevallia popayanensis	Dryadella simula
Masdevallia popowiana Königer & J.G.Weinm.bis*	
Masdevallia popowiana Königer & J.Portilla	Masdevallia bicornis

*For explanation see page 2, point 6
*Voir les explications page 10, point 6
*Para mayor explicación, véase la página 20, point 6

Part I: All Names / Tous les Noms / Todos los Nombres

ALL NAMES	ACCEPTED NAMES
Masdevallia porcelliceps	**Masdevallia macroglossa**
Masdevallia porphyrea	
Masdevallia portillae	
Masdevallia posadae	
Masdevallia pozoi	
Masdevallia princeps	
Masdevallia proboscoidea	
Masdevallia prodigiosa	
Masdevallia prolixa	
Masdevallia prosartema	
Masdevallia pseudominuta	**Masdevallia kyphonantha**
Masdevallia psittacina	**Dracula psittacina**
Masdevallia psyche	**Dracula psyche**
Masdevallia pteroglossa	
Masdevallia pterygiophora	**Diodonopsis pterygiophora**
Masdevallia pulcherrima	
Masdevallia pulex	**Pleurothallis macroblepharis**
Masdevallia pulvinaris	**Scaphosepalum pulvinare**
Masdevallia pumila	
Masdevallia punctata	**Scaphosepalum anchoriferum**
Masdevallia purpurella	
Masdevallia purpurina	**Masdevallia amabilis**
Masdevallia pusilla	**Dracula pusilla**
Masdevallia pusiola	**Dryadella pusiola**
Masdevallia pygmaea	**Diodonopsis pygmaea**
Masdevallia pyknosepala	
Masdevallia pyxis	
Masdevallia quasimodo	
Masdevallia quilichaoënsis	**Dracula iricolor**
Masdevallia racemosa	
Masdevallia radiosa	**Dracula radiosa**
Masdevallia rafaeliana	
Masdevallia rana-aurea	**Masdevallia os-viperae**
Masdevallia rauhii	**Masdevallia mezae**
Masdevallia receptrix	
Masdevallia rechingeriana	
Masdevallia recurvata	
Masdevallia reflexa Misas non Schltr.	**Masdevallia misasii***
Masdevallia reflexa Schltr.	**Masdevallia cupularis**
Masdevallia regina	
Masdevallia reichenbachiana	
Masdevallia remotiflora	**Masdevallia amanda**
Masdevallia renzii	
Masdevallia repanda	
Masdevallia replicata	
Masdevallia restrepioidea	**Masdevallia fasciata**
Masdevallia revoluta	
Masdevallia rex	
Masdevallia rhinophora	
Masdevallia rhodehameliana	
Masdevallia rhopalura	**Masdevallia molossoides**
Masdevallia richardsoniana	
Masdevallia richteri	**Masdevallia vargasii**
Masdevallia ricii	
Masdevallia rigens	
Masdevallia rimarima-alba	
Masdevallia robusta	
Masdevallia rodolfoi	

*For explanation see page 2, point 6
*Voir les explications page 10, point 6
54   *Para mayor explicación, véase la página 20, point 6

ALL NAMES	ACCEPTED NAMES
Masdevallia rodrigueziana	**Masdevallia wendlandiana**
Masdevallia roezlii	**Dracula roezlii**
Masdevallia rolandorum	
Masdevallia rolfeana	
Masdevallia rosea	
Masdevallia roseola	
Masdevallia rubeola	
Masdevallia rubiginosa	
Masdevallia rufescens	
Masdevallia rufolutea	**Masdevallia picea**
Masdevallia rugulosa	
Masdevallia saltatrix	
Masdevallia sanchezii	
Masdevallia sanctae-fidei	
Masdevallia sanctae-inesiae	
Masdevallia sanctae-rosae	
Masdevallia sanguinea	
Masdevallia santiagoörum	**Masdevallia aurea**
Masdevallia sarcophylla	**Masdevallia campyloglossa**
Masdevallia saulii	**Masdevallia fuchsii**
Masdevallia scabrilinguis	
Masdevallia scalpellifera	
Masdevallia scandens	
Masdevallia scapha	**Masdevallia navicularis**
Masdevallia sceptrum	
Masdevallia schildhaueri	**Masdevallia ensata**
Masdevallia schizantha	
Masdevallia schizopetala	
Masdevallia schizostigma	
Masdevallia schlimii	
Masdevallia schmidtchenii	**Masdevallia molossus**
Masdevallia schmidt-mummii	
Masdevallia schoonenii	
Masdevallia schroederae	**Masdevallia schroederiana**
Masdevallia schroederiana	
Masdevallia schudelii	
Masdevallia scitula	
Masdevallia scobina	
Masdevallia scopaea	
Masdevallia segrex	
Masdevallia segurae	
Masdevallia selenites	
Masdevallia semiteres	
Masdevallia × **senghasiana**	
Masdevallia serendipita	
Masdevallia sernae	
Masdevallia sertula	
Masdevallia sessilis	**Dryadella aviceps**
Masdevallia setacea	
Masdevallia setipes	
Masdevallia severa	**Dracula severa**
Masdevallia shiraishii	**Masdevallia harlequina**
Masdevallia shuttleworthii	**Masdevallia caudata**
Masdevallia sijmiana	**Masdevallia posadae**
Masdevallia simia	**Dracula simia**
Masdevallia simula	**Dryadella simula**
Masdevallia simulatrix	**Dryadella simulatrix**
Masdevallia siphonantha	

*For explanation see page 2, point 6
*Voir les explications page 10, point 6
*Para mayor explicación, véase la página 20, point 6   **55**

ALL NAMES	ACCEPTED NAMES
Masdevallia smallmaniana	
Masdevallia sodiroi	**Dracula sodiroi**
Masdevallia soennemarkii	
Masdevallia solomonii	
Masdevallia sororcula	**Masdevallia mooreana**
Masdevallia spathulifolia	**Masdevallia polysticta**
Masdevallia spectrum	**Dracula severa**
Masdevallia sphenopetala	**Masdevallia corazonica**
Masdevallia spilantha	
Masdevallia × splendida	
Masdevallia sprucei	
Masdevallia staaliana	
Masdevallia stenantha	**Masdevallia tubulosa**
Masdevallia stenorhynchos	
Masdevallia stercorea	**Masdevallia rigens**
Masdevallia stigii	
Masdevallia stirpis	
Masdevallia strattoniana	
Masdevallia striatella	
Masdevallia strigosa	
Masdevallia strobelii	
Masdevallia × strumella	
Masdevallia strumifera	
Masdevallia strumosa	
Masdevallia stumpflei	
Masdevallia subumbellata	**Masdevallia bicolor**
Masdevallia suinii	
Masdevallia sulphurea	**Masdevallia bonplandii**
Masdevallia sulphurella	
Masdevallia sumapazensis	
Masdevallia summersii	**Dryadella summersii**
Masdevallia superflua	**Masdevallia striatella**
Masdevallia surinamensis	**Masdevallia minuta**
Masdevallia susanae	**Dryadella susanae**
Masdevallia swertiaefolia	**Scaphosepalum swertiaefolium**
Masdevallia × synthesis	
Masdevallia syringodes	**Masdevallia tubulosa**
Masdevallia tarantula	**Dracula tubeana**
Masdevallia teaguei	
Masdevallia tentaculata	
Masdevallia tenuicaudata	**Masdevallia nidifica**
Masdevallia tenuipes	**Masdevallia herradurae**
Masdevallia tenuissima	**Trisetella tenuissima**
Masdevallia terborchii	
Masdevallia theleura	
Masdevallia thienii	
Masdevallia tinekeae	
Masdevallia titan	
Masdevallia tokachiorum	
Masdevallia tonduzii	
Masdevallia torta	
Masdevallia torulosa	**Masdevallia carruthersiana**
Masdevallia tovarensis	
Masdevallia trautmanniana	
Masdevallia trechsliniana	**Masdevallia bryophila**
Masdevallia triangularis	
Masdevallia triaristella	**Trisetella triaristella**
Masdevallia tricallosa	

*For explanation see page 2, point 6
*Voir les explications page 10, point 6
56   *Para mayor explicación, véase la página 20, point 6

ALL NAMES	ACCEPTED NAMES
Masdevallia tricarinata	**Pleurothallis tricarinata**
Masdevallia triceratops	**Dracula mopsus**
Masdevallia trichaete	**Trisetella triglochin**
Masdevallia trichroma	**Dracula iricolor**
Masdevallia tricolor hort. non Rchb.f.	**Masdevallia coccinea**
Masdevallia tricolor Rchb.f. (sphalm.)	**Dracula iricolor**
Masdevallia tricolor Rchb.f.	**Masdevallia caudata**
Masdevallia tricycla	
Masdevallia tridactylites	**Trisetella triglochin**
Masdevallia tridens	
Masdevallia tridentata	**Masdevallia infracta**
Masdevallia trifurcata	
Masdevallia triglochin	**Trisetella triglochin**
Masdevallia trigonopetala	
Masdevallia trinema Rchb.f.	**Dracula velutina**
Masdevallia trinema sensu Woolward & F.Lehm. non Rchb.f.	**Dracula platycrater**
Masdevallia trinemoides	**Masdevallia fasciata**
Masdevallia trionyx	**Masdevallia falcago**
Masdevallia trioön	**Masdevallia bangii**
Masdevallia triquetra	**Masdevallia infracta**
Masdevallia triseta	**Trisetella triglochin**
Masdevallia trivenia	**Masdevallia paquishae**
Masdevallia trochilus	
Masdevallia troglodytes	**Dracula benedictii**
Masdevallia truncata	
Masdevallia tsubotae	
Masdevallia tubata	
Masdevallia tubeana	**Dracula tubeana**
Masdevallia tubuliflora	
Masdevallia tubulosa	
Masdevallia tuerckheimii	**Masdevallia floribunda**
Masdevallia ulei	**Masdevallia wendlandiana**
Masdevallia uncifera	
Masdevallia uniflora Kunth non Ruiz & Pav.	**Masdevallia bonplandii**
Masdevallia uniflora Ruiz & Pav.*	
Masdevallia urceolaris	
Masdevallia urosalpinx	**Masdevallia constricta**
Masdevallia urostachya	**Masdevallia sceptrum**
Masdevallia ustulata	
Masdevallia utriculata	
Masdevallia valenciae	
Masdevallia valifera	**Masdevallia velifera**
Masdevallia vampira	**Dracula vampira**
Masdevallia vargasii	
Masdevallia vasquezii	
Masdevallia veitchiana	
Masdevallia velella	
Masdevallia velifera	
Masdevallia velox	**Masdevallia dalessandroi**
Masdevallia velutina	**Dracula velutina**
Masdevallia venatoria	
Masdevallia venezuelana	
Masdevallia venosa	**Dracula venosa**
Masdevallia ventricosa	
Masdevallia ventricularia	
Masdevallia venus	
Masdevallia venusta	
Masdevallia verecunda	

*For explanation see page 2, point 6
*Voir les explications page 10, point 6
*Para mayor explicación, véase la página 20, point 6

Part I: All Names / Tous les Noms / Todos los Nombres

ALL NAMES	ACCEPTED NAMES
Masdevallia verrucosa	Scaphosepalum verrucosum
Masdevallia versicolor	Masdevallia coccinea
Masdevallia vespertilio	Dracula vespertilio
Masdevallia vexillifera	
Masdevallia vidua	
Masdevallia vieirana	
Masdevallia vilifera	Masdevallia velifera
Masdevallia villegasii	
Masdevallia virens	
Masdevallia virgo-cuencae	
Masdevallia vittata	Trisetella vittata
Masdevallia vittatula	
Masdevallia vomeris	
Masdevallia vulcanica	Diodonopsis anachaeta
Masdevallia wageneriana	
Masdevallia wallisii	Dracula wallisii
Masdevallia walteri	
Masdevallia weberbaueri	
Masdevallia welischii	
Masdevallia wendlandiana	
Masdevallia whiteana	
Masdevallia winniana	Dracula roezlii
Masdevallia woolwardiae	Dracula woolwardiae
Masdevallia × wubbenii	
Masdevallia wuelfinghoffiana	
Masdevallia wuerstlei	
Masdevallia wurdackii	
Masdevallia xanthina	
Masdevallia xanthodactyla	
Masdevallia xanthura	Masdevallia bicolor
Masdevallia xerophila	Masdevallia pteroglossa
Masdevallia ximenae	
Masdevallia xipheres	Porroglossum muscosum
Masdevallia xiphium	Masdevallia ensata
Masdevallia xylina	
Masdevallia yauaperyensis	Masdevallia wendlandiana
Masdevallia yungasensis	
Masdevallia zahlbruckneri	
Masdevallia zamorensis	
Masdevallia zapatae	
Masdevallia zebracea	
Masdevallia zebrina	Dryadella zebrina
Masdevallia zongoënsis	
Masdevallia zumbae	
Masdevallia zumbuehlerae	
Masdevallia zygia	
Orxera cornuta	Aerides odorata
Panisea bilamellata	Coelogyne bilamellata
Pholidota suaveolens	Coelogyne suaveolens
Physosiphon bangii	Masdevallia bangii
Physosiphon lansbergii	Masdevallia lansbergii
Pleione anceps	Coelogyne anceps
Pleione asperata	Coelogyne asperata
Pleione barbata	Coelogyne barbata
Pleione bilamellata	Coelogyne bilamellata
Pleione brachyptera	Coelogyne brachyptera
Pleione brevifolia	Coelogyne punctulata
Pleione breviscapa	Coelogyne breviscapa

*For explanation see page 2, point 6
*Voir les explications page 10, point 6
*Para mayor explicación, véase la página 20, point 6

ALL NAMES	ACCEPTED NAMES
Pleione chinensis Kuntze	Coelogyne fimbriata
Pleione corrugata	Coelogyne nervosa
Pleione corymbosa	Coelogyne corymbosa
Pleione cumingii	Coelogyne cumingii
Pleione cycnoches	Coelogyne fuscescens
Pleione dayana	Coelogyne pulverula
Pleione elata	Coelogyne stricta
Pleione fimbriata	Coelogyne fimbriata
Pleione flaccida	Coelogyne flaccida
Pleione flavida	Coelogyne prolifera
Pleione foerstermannii	Coelogyne foerstermannii
Pleione fuliginosa	Coelogyne fuliginosa
Pleione fuscescens	Coelogyne fuscescens
Pleione glandulosa	Coelogyne glandulosa
Pleione goweri	Coelogyne punctulata
Pleione graminifolia	Coelogyne viscosa
Pleione griffithii	Coelogyne griffithii
Pleione huettneriana	Coelogyne huettneriana
Pleione incrassata	Coelogyne incrassata
Pleione lactea	Coelogyne flaccida
Pleione lauterbachiana	Coelogyne miniata
Pleione lentiginosa	Coelogyne lentiginosa
Pleione longebracteata	Coelogyne cumingii
Pleione longifolia	Coelogyne longifolia
Pleione longipes	Coelogyne longipes
Pleione macrobulbon	Coelogyne rochussenii
Pleione maingayi	Coelogyne foerstermannii
Pleione massangeana	Coelogyne tomentosa
Pleione micrantha	Coelogyne micrantha
Pleione miniata	Coelogyne miniata
Pleione nervosa	Coelogyne nervosa
Pleione nitida	Coelogyne punctulata
Pleione occultata	Coelogyne occultata
Pleione ochracea	Coelogyne nitida
Pleione odoratissima	Coelogyne odoratissima
Pleione pandurata	Coelogyne pandurata
Pleione parishii	Coelogyne parishii
Pleione plantaginea	Coelogyne rochussenii
Pleione prolifera	Coelogyne prolifera
Pleione psittacina	Coelogyne rumphii
Pleione rigida	Coelogyne rigida
Pleione rochussenii	Coelogyne rochussenii
Pleione rossiana	Coelogyne trinervis
Plcionc rumphii	Coelogyne rumphii
Pleione sanderiana	Coelogyne sanderiana
Pleione schilleriana	Coelogyne schilleriana
Pleione simplex	Coelogyne miniata
Pleione sparsa	Coelogyne sparsa
Pleione speciosa	Coelogyne speciosa
Pleione speciosissima	Coelogyne cristata
Pleione stenochila	Coelogyne stenochila
Pleione suaveolens	Coelogyne suaveolens
Pleione testacea	Coelogyne testacea
Pleione tomentosa	Coelogyne tomentosa
Pleione trinervis	Coelogyne trinervis
Pleione triplicata	Coelogyne fuliginosa
Pleione ustulata	Coelogyne ustulata

*For explanation see page 2, point 6
*Voir les explications page 10, point 6
*Para mayor explicación, véase la página 20, point 6

Part I: All Names / Tous les Noms / Todos los Nombres

ALL NAMES	ACCEPTED NAMES
Pleione viscosa	**Coelogyne viscosa**
Pleurothallis plantaginea	**Masdevallia plantaginea**
Pleurothallis sanctae-rosae	**Masdevallia sanctae-rosae**
Polytoma odorifera	**Aerides odorata**
Portillia popowiana	**Masdevallia bicornis**
Ptychogyne bimaculata	**Coelogyne flexuosa**
Ptychogyne flexuosa	**Coelogyne flexuosa**
Rodrigoa bilabiata	**Masdevallia platyglossa**
Rodrigoa cryptocopis	**Masdevallia picturata**
Rodrigoa diversifolia	**Masdevallia parvula**
Rodrigoa fasciata	**Masdevallia fasciata**
Rodrigoa heteroptera	**Masdevallia heteroptera**
Rodrigoa meleagris	**Masdevallia meleagris**
Rodrigoa segurae	**Masdevallia segurae**
Saccolabium huttonii	**Aerides thibautiana**
Saccolabium lineare	**Aerides ringens**
Saccolabium paniculatum	**Aerides ringens**
Saccolabium ringens	**Aerides ringens**
Saccolabium rubescens	**Aerides rubescens**
Saccolabium speciosum	**Aerides maculosa**
Saccolabium wightianum	**Aerides ringens**
Scaphosepalum panamense	**Masdevallia livingstoneana**
Specklinia plantaginea	**Masdevallia plantaginea**
Vanda flabellata	**Aerides flabellata**

*For explanation see page 2, point 6
*Voir les explications page 10, point 6
60
*Para mayor explicación, véase la página 20, point 6

PART II: ACCEPTED NAMES IN CURRENT USE
Ordered alphabetically on Accepted Names for the genera:

Aerides, *Coelogyne*, *Comparettia* and *Masdevallia*

DEUXIÈME PARTIE: NOMS ACCEPTES D'USAGE COURANT
Par ordre alphabétique des noms acceptés pour les genre:

Aerides, *Coelogyne*, *Comparettia* et *Masdevallia*

PARTE II: NOMBRES ACEPTADOS DE USO ACTUAL
Presentados por orden alfabético: nombres aceptados para el genero:

Aerides, *Coelogyne*, *Comparettia* y *Masdevallia*

AERIDES BINOMIALS IN CURRENT USE

AERIDES BINOMES ACTUELLEMENT EN USAGE

AERIDES BINOMIALES UTILIZADOS NORMALMENTE

Aerides augustiana Rolfe
Aerides augustianum Rolfe

Distribution: Philippines (the)

Aerides crassifolia C.S.P.Parish ex Burb.
Aerides crassifolium C.S.P.Parish ex Burb.
Aerides expansa Rchb.f.
Aerides expansum Rchb.f.

Distribution: India, Lao People's Democratic Republic (the), Myanmar, Thailand, Viet Nam

Aerides crispa Lindl.
Aerides brockessii Heynh.
Aerides brookei Bateman ex Lindl.
Aerides crispum Lindl.
Aerides lindleyana Wight
Aerides lindleyanum Wight
Aerides warneri Hook.f.

Distribution: India

Aerides emericii Rchb.f.

Distribution: India

Aerides falcata Lindl. & Paxton
Aerides falcatum Lindl. & Paxton
Aerides larpentae Rchb.f.
Aerides mendelii E.Morren
Aerides retrofracta Wall. ex Hook.f. nom.nud.
Aerides retrofractum Wall. ex Hook.f. nom.nud.
Aerides siamense Klinge
Aerides siamensis Klinge

Distribution: Cambodia, India, Indonesia, Lao People's Democratic Republic (the), Malaysia, Myanmar, Thailand, Viet Nam

Aerides flabellata Rolfe ex Downie
Aerides flabellatum Rolfe ex Downie
Vanda flabellata (Rolfe ex Downie) Christenson

Distribution: China, Lao People's Democratic Republic (the), Myanmar, Thailand

Part II: Aerides

Aerides houlletiana Rchb.f.
Aerides ellisii J.Anderson
Aerides houlletianum Rchb.f.
Aerides picotiana hort. ex Rchb.f.
Aerides picotianum hort. ex Rchb.f.
Aerides platychila Rolfe
Aerides platychilum Rolfe

Distribution: Cambodia, Lao People's Democratic Republic (the), Myanmar, Thailand, Viet Nam

Aerides inflexa Teijsm. & Binn.
Aerides bernhardiana Rchb.f.
Aerides bernhardianum Rchb.f.
Aerides inflexum Teijsm. & Binn.

Distribution: Indonesia

Aerides × jansonii Rolfe

Distribution: Myanmar

Aerides krabiensis Seidenf.
Aerides krabiense Seidenf.
Aerides multiflora auct. non Roxb.
Aerides multiflorum auct. non Roxb.

Distribution: Malaysia, Thailand

Aerides lawrenceae Rchb.f.
Aerides lawrenciae Rchb.f.
Aerides sanderiana Rchb.f.
Aerides sanderianum Rchb.f.

Distribution: Philippines (the)

Aerides leeana Rchb.f.
Aerides jarckiana Schltr.
Aerides jarckianum Schltr.
Aerides leeanum Rchb.f.
Aerides recurvipes J.J.Sm.

Distribution: Philippines (the)

Aerides maculosa Lindl.
Aerides illustre Rchb.f.
Aerides illustris Rchb.f.
Aerides maculosum Lindl.
Aerides margaritacea hort. nom.nud.
Aerides margaritaceum hort. nom.nud.
Aerides schroederi hort. ex Rchb.f.
Gastrochilus speciosus (Wight) Kuntze

Saccolabium speciosum Wight

Distribution: India

Aerides mcmorlandii B.S.Williams

Distribution: India

Aerides multiflora Roxb.
Aerides affine Wall. ex Lindl.
Aerides affinis Wall. ex Lindl.
Aerides godefroyana Rchb.f.
Aerides godefroyanum Rchb.f.
Aerides lobbii Lem. non Teijsm. & Binn.
Aerides multiflorum Roxb.
Aerides veitchii hort. ex B.S.Williams
Cleisostoma vacherotiana Guillaumin
Epidendrum geniculatum Hook.f.

Distribution: Bangladesh, Bhutan, Cambodia, India, Lao People's Democratic Republic
(the), Malaysia, Myanmar, Nepal, Thailand, Viet Nam

Aerides odorata Lour.
Aeeridium odorum Salisb.
Aerides ballantiniana Rchb.f.
Aerides ballantinianum Rchb.f.
Aerides cornuta Roxb.
Aerides cornutum Roxb.
Aerides dayana hort. ex Guillaumin
Aerides dayanum hort. ex Guillaumin
Aerides duquesnei Regnier ex Costantin
Aerides flavida Lindl.
Aerides flavidum Lindl.
Aerides jucunda Rchb.f.
Aerides jucundum Rchb.f.
Aerides latifolia (Thunb. ex Sw.) Sw.
Aerides latifolium (Thunb. ex Sw.) Sw.
Aerides micholitzii Rolfe
Aerides nobile R.Warner
Aerides nobilis R.Warner
Aerides odorata Reinw. ex non Lour.
Aerides odoratum Lour.
Aerides odoratum Reinw. ex non Lour.
Aerides reichenbachii Linden
Aerides rohaniana Rchb.f.
Aerides rohanianum Rchb.f.
Aerides suaveolens Blume. non Roxb.
Aerides suavissima Lindl.
Aerides suavissimum Lindl.
Aerides virens Lindl.
Aerides wilsoniana hort.
Aerides wilsonianum hort.
Epidendrum. aerides Raeusch.
Epidendrum odoratum Poir.
Limodorum latifolium Thunb. ex Sw.
Orxera cornuta (Roxb.) Raf.

Part II: Aerides

Polytoma odorifera Lour. ex Gomes

Distribution: Bhutan, Cambodia, China, India, Indonesia, Lao People's Democratic Republic (the), Malaysia, Myanmar, Nepal, Philippines (the), Thailand, Viet Nam

Aerides quinquevulnera Lindl.
Aerides alba Sander ex Stein
Aerides album Sander ex Stein
Aerides farmeri Boxall ex Náves
Aerides fenzliana Rchb.f.
Aerides fenzlianum Rchb.f.
Aerides maculata Llanos non Buch.-Ham. ex Sm.
Aerides maculatum Llanos non Buch.-Ham. ex Sm.
Aerides marginata Rchb.f.
Aerides marginatum Rchb.f.
Aerides ortgiesiana Rchb.f.
Aerides ortgiesianum Rchb.f.
Aerides quinquevulnerum Lindl.
Aerides reversa J.J.Sm.
Aerides reversum J.J.Sm.
Aerides roebelenii Rchb.f.
Aerides savageana Sander ex H.J.Veitch
Aerides savageanum Sander ex H.J.Veitch

Distribution: Papua New Guinea, Philippines (the)

Aerides ringens (Lindl.) C.E.C.Fisch.
Aerides lineare Hook.f.
Aerides linearis Hook.f.
Aerides radicosa A.Rich.
Aerides radicosum A.Rich.
Cymbidium lineare Heyne ex Wall.
Gastrochilus ringens (Lindl.) Kuntze
Saccolabium lineare Lindl.
Saccolabium paniculatum Wight
Saccolabium ringens Lindl.
Saccolabium wightianum Lindl.

Distribution: India, Sri Lanka

Aerides rosea Lodd. ex Lindl. & Paxton
Aerides fieldingii Lodd. ex E.Morren
Aerides roseum Lodd. ex Lindl. & Paxton
Aerides trigona Klotzsch
Aerides trigonum Klotzsch
Aerides williamsii R.Warner

Distribution: Bhutan, China, India, Lao People's Democratic Republic (the), Myanmar, Thailand, Viet Nam

Aerides rubescens (Rolfe) Schltr.
Ascocentrum rubescens (Rolfe) P.F.Hunt
Saccolabium rubescens Rolfe

Distribution: Viet Nam

Aerides sukauensis Shim
Aerides sukauense Shim

Distribution: Malaysia

Aerides thibautiana Rchb.f.
Aerides huttonii (Hook.f.) H.J.Veitch
Aerides shibatiana Boxall ex Náves (sphalm.)
Aerides shibatianum Boxall ex Náves (sphalm.)
Aerides thibautianum Rchb.f.
Saccolabium huttonii Hook.f.

Distribution: Indonesia, Philippines (the)

Aerides timorana Miq.
Aerides pallida non Roxb. nec (Blume) Lindl.
Aerides pallidum non Roxb. nec (Blume) Lindl.
Aerides timoranum Miq.

Distribution: Indonesia

COELOGYNE BINOMIALS IN CURRENT USE

COELOGYNE BINOMES ACTUELLEMENT EN USAGE

COELOGYNE BINOMIALES UTILIZADOS NORMALMENTE

Coelogyne acutilabium de Vogel

Distribution: Malaysia

Coelogyne albobrunnea J.J.Sm.

Distribution: Indonesia, Malaysia

Coelogyne albolutea Rolfe

Distribution: India

Coelogyne anceps Hook.f.
  Pleione anceps (Hook.f.) Kuntze

Distribution: Malaysia

Coelogyne asperata Lindl.
  Coelogyne edelfeldtii F.Muell. & Kraenzl.
  Coelogyne lowii Paxton
  Coelogyne macrophylla Teijsm. & Binn.
  Coelogyne pustulosa Ridl.
  Cymbidium robustum Gilli
  Pleione asperata (Lindl.) Kuntze

Distribution: Brunei Darussalam, Indonesia, Malaysia, Papua New Guinea, Philippines (the), Solomon Islands

Coelogyne assamica Linden & Rchb.f.
  Coelogyne annamensis (Lindl. & Rchb.f.) Rolfe
  Coelogyne dalatensis Gagnep.
  Coelogyne fuscescens Hook.f. non Lindl.
  Coelogyne saigonensis Gagnep.
  Coelogyne siamensis Rolfe
  Cymbidium evrardii Guillaumin

Distribution: Bhutan, China, India, Lao People's Democratic Republic (the), Myanmar, Thailand, Viet Nam

Coelogyne barbata Griff.
  Pleione barbata (Lindl. ex Griff.) Kuntze

Distribution: Bhutan, China, India, Myanmar, Nepal

Part II: Coelogyne

Coelogyne beccarii Rchb.f.
Coelogyne micholitziana Kraenzl.

Distribution: Indonesia, Papua New Guinea, Solomon Islands

Coelogyne bicamerata J.J.Sm.

Distribution: Indonesia

Coelogyne bilamellata Lindl.
Panisea bilamellata (Lindl.) Rchb.f.
Pleione bilamellata (Lindl.) Kuntze

Distribution: Philippines (the),

Coelogyne borneensis Rolfe

Distribution: Indonesia

Coelogyne brachygyne J.J.Sm.

Distribution: Indonesia

Coelogyne brachyptera Rchb.f.
Pleione brachyptera (Rchb.f.) Kuntze

Distribution: Myanmar, Viet Nam

Coelogyne breviscapa Lindl.
Coelogyne angustifolia Wight non A.Rich.
Pleione breviscapa (Lindl.) Kuntze

Distribution: India, Sri Lanka

Coelogyne bruneiensis de Vogel

Distribution: Brunei Darussalam

Coelogyne buennemeyeri J.J.Sm.

Distribution: Indonesia

Coelogyne calcarata J.J.Sm.

Distribution: Indonesia

Coelogyne calcicola Kerr

Distribution: China, Lao People's Democratic Republic (the), Myanmar, Thailand, Viet Nam

Coelogyne caloglossa Schltr.

Distribution: Indonesia

Coelogyne candoonensis Ames

Distribution: Philippines (the)

Coelogyne carinata Rolfe
Coelogyne alata A.Millar
Coelogyne oligantha Schltr.
Coelogyne sarasinorum Kraenzl.
Coelogyne truncicola Schltr.

Distribution: Indonesia, Papua New Guinea, Solomon Islands

Coelogyne celebensis J.J.Sm.
Coelogyne platyphylla Schltr.

Distribution: Indonesia

Coelogyne chanii Gravend. & de Vogel

Distribution: Malaysia

Coelogyne chlorophaea Schltr.

Distribution: Indonesia

Coelogyne chloroptera Rchb.f.

Distribution: Philippines (the)

Coelogyne chrysotropis Schltr.

Distribution: Indonesia

Coelogyne clemensii Ames & C.Schweinf.

Distribution: Indonesia, Malaysia

Part II: Coelogyne

Coelogyne compressicaulis Ames & C.Schweinf.

Distribution: Indonesia, Malaysia

Coelogyne concinna Ridl.

Distribution: Indonesia

Coelogyne confusa Ames

Distribution: Philippines (the)

Coelogyne contractipetala J.J.Sm.

Distribution: Indonesia

Coelogyne corymbosa Lindl.
  Pleione corymbosa (Lindl.) Kuntze

Distribution: Bhutan, China, India, Nepal

Coelogyne crassiloba J.J.Sm.

Distribution: Indonesia, Malaysia

Coelogyne craticulaelabris Carr

Distribution: Brunei Darussalam, Indonesia, Malaysia

Coelogyne cristata Lindl.
  Cymbidium speciosissimum D.Don
  Pleione speciosissima (Lindl.) Kuntze

Distribution: Bhutan, China, India, Nepal, Thailand

Coelogyne cumingii Lindl.
  Coelogyne casta Ridl.
  Coelogyne longibracteata Hook.f.
  Pleione cumingii (Lindl.) Kuntze
  Pleione longebracteata (Hook.f.) Kuntze

Distribution: Indonesia, Lao People's Democratic Republic (the), Malaysia, Singapore, Thailand

Coelogyne cuprea H.Wendl. & Kraenzl.

Distribution: Indonesia, Malaysia

74

Coelogyne dichroantha Gagnep.

Distribution: Viet Nam

Coelogyne distans J.J.Sm.

Distribution: Indonesia, Malaysia

Coelogyne dulitensis Carr

Distribution: Malaysia

Coelogyne eberhardtii Gagnep.

Distribution: Viet Nam

Coelogyne ecarinata C.Schweinf.

Distribution: Myanmar

Coelogyne echinolabium de Vogel

Distribution: Brunei Darussalam, Indonesia, Malaysia

Coelogyne elmeri Ames

Distribution: Philippines (the)

Coelogyne endertii J.J.Sm.

Distribution: Indonesia, Malaysia

Coelogyne esquirolii Schltr.
Coelogyne esquirolei Schltr.

Distribution: China

Coelogyne exalata Ridl.
 Coelogyne subintegra J.J.Sm.

Distribution: Brunei Darussalam, Malaysia

Coelogyne filipeda Gagnep.
 Coelogyne longiana Aver.
 Coelogyne prolifera sensu Gagnep. non Lindl.

Distribution: Viet Nam

Part II: Coelogyne

Coelogyne fimbriata Lindl.
Coelogyne arunachalensis H.J.Chowdhery & G.D.Pal
Coelogyne laotica Gagnep.
Coelogyne leungiana S.Y.Hu
Coelogyne ovalis sensu Pfitzer & Kraenzl. non Lindl.
Coelogyne primulina Barretto
Coelogyne xerophyta Hand.-Mazz.
Pleione chinense Kraenzl.
Pleione fimbriata (Lindl.) Kraenzl.

Distribution: Bhutan, Cambodia, China, India, Lao People's Democratic Republic (the), Malaysia, Myanmar, Nepal, Thailand, Viet Nam

Coelogyne flaccida Lindl.
Coelogyne huettneriana sensu Hook.f. non Rchb.f.
Coelogyne lactea Rchb.f.
Pleione flaccida (Lindl.) Kuntze
Pleione lactea (Rchb.f.) Kuntze

Distribution: China, India, Lao People's Democratic Republic (the), Myanmar, Nepal, Thailand, Viet Nam

Coelogyne flexuosa Rolfe
Coelogyne bimaculata Ridl.
Ptychogyne bimaculata (Ridl.) Pfitzer
Ptychogyne flexuosa (Rolfe) Pfitzer

Distribution: Indonesia, Malaysia, Singapore

Coelogyne foerstermannii Rchb.f.
Coelogyne kingii Hook.f.
Coelogyne maingayi Hook.f.
Pleione foerstermannii (Rchb.f.) Kuntze
Pleione maingayi (Hook.f.) Kuntze

Distribution: Indonesia, Malaysia, Singapore

Coelogyne fonstenebrarum P.O'Byrne

Distribution: Indonesia

Coelogyne formosa Schltr.

Distribution: Indonesia

Coelogyne fragrans Schltr.

Distribution: Indonesia, Papua New Guinea

Coelogyne fuerstenbergiana Schltr.

Distribution: Indonesia

Coelogyne fuliginosa Lodd. ex Hook.
Coelogyne fimbriata auct. non Lindl.
Pleione fuliginosa (Hook.f.) Kuntze
Pleione triplicata (Rchb.f.) Kuntze

Distribution: India, Indonesia, Myanmar

Coelogyne fuscescens Lindl.
Coelogyne brunnea Lindl.
Coelogyne cycnoches C.S.P.Parish & Rchb.f.
Pleione cycnoches (C.S.P.Parish & Rchb.f.) Kuntze
Coelogyne integrilabia (Pfitzer) Schltr.
Pleione fuscescens (Lindl.) Kuntze

Distribution: Bhutan, China, India, Lao People's Democratic Republic (the), Myanmar, Nepal, Thailand, Viet Nam

Coelogyne genuflexa Ames & C.Schweinf.
Coelogyne reflexa J.J.Wood & C.L.Chan

Distribution: Brunei Darussalam, Indonesia, Malaysia

Coelogyne ghatakii T.K. Paul, S.K. Basu & M.C. Biswas

Distribution: India

Coelogyne gibbifera J.J.Sm.
Coelogyne macroloba J.J.Sm.

Distribution: Brunei Darussalam, Indonesia, Malaysia

Coelogyne glandulosa Lindl.
Pleione glandulosa (Lindl.) Kuntze

Distribution: India

Coelogyne gongshanensis H.Li ex S.C.Chen

Distribution: China

Coelogyne griffithii Hook.f.
Pleione griffithii (Hook.f.) Kuntze

Distribution: China, India, Myanmar, Viet Nam

Part II: Coelogyne

Coelogyne guamensis Ames
Coelogyne palawensis Tuyama

Distribution: Guam (Dependant Territory of the United States of America), Commonwealth of the Northern Mariana Islands (Dependant Territory of the United States of America), Palau

Coelogyne hajrae Phukan

Distribution: India

Coelogyne harana J.J.Sm.

Distribution: Indonesia, Malaysia

Coelogyne hirtella J.J.Sm.
Coelogyne radioferens sensu J.J.Sm. non Ames & C.Schweinf.
Coelogyne radiosa J.J.Sm.

Distribution: Brunei Darussalam, Indonesia, Malaysia

Coelogyne hitendrae S.Das & S.K.Jain

Distribution: India

Coelogyne holochila P.F.Hunt & Summerh.
Coelogyne elata sensu Hook. non Lindl.

Distribution: Bhutan, China, India, Myanmar, Nepal

Coelogyne huettneriana Rchb.f.
Pleione huettneriana (Rchb.f.) Kuntze

Distribution: Myanmar, Thailand, Viet Nam

Coelogyne imbricans J.J.Sm.

Distribution: Indonesia, Malaysia

Coelogyne incrassata (Blume) Lindl.
Chelonanthera incrassata Blume
Pleione incrassata (Blume) Kuntze

Distribution: Brunei Darussalam, Indonesia, Malaysia

Coelogyne integerrima Ames

Distribution: Philippines (the)

Coelogyne integra Schltr.

Distribution: Indonesia, Philippines (the)

Coelogyne judithiae P.Taylor

Distribution: Malaysia

Coelogyne kaliana P.J.Cribb
  Coelogyne massangeana auct. non Rchb.f.

Distribution: Malaysia

Coelogyne kelamensis J.J.Sm.

Distribution: Indonesia, Malaysia

Coelogyne kemiriensis J.J.Sm.

Distribution: Indonesia

Coelogyne kinabaluensis Ames & C.Schweinf.

Distribution: Indonesia, Malaysia

Coelogyne lacinulosa J.J.Sm.

Distribution: Indonesia

Coelogyne latiloba de Vogel

Distribution: Indonesia, Malaysia

Coelogyne lawrenceana Rolfe
  Coelogyne fleuryi Gagnep.

Distribution: Viet Nam

Coelogyne lentiginosa Lindl.
  Pleione lentiginosa (Lindl.) Kuntze

Distribution: Myanmar, Thailand, Viet Nam

Coelogyne leucantha W.W.Sm.

Distribution: China, Myanmar

Part II: Coelogyne

Coelogyne lockii Aver.

Distribution: China, Viet Nam

Coelogyne loheri Rolfe

Distribution: Philippines (the)

Coelogyne longeciliata Teijsm. & Binn.

Distribution: India

Coelogyne longibulbosa Ames & C.Schweinf.

Distribution: Indonesia, Malaysia

Coelogyne longifolia (Blume) Lindl.
   Chelonanthera longifolia Blume
   Cymbidium stenopetalum Reinw. ex Lindl.
   Pleione longifolia (Blume) Kuntze

Distribution: Indonesia, Myanmar

Coelogyne longipes Lindl.
   Pleione longipes (Lindl.) Kuntze

Distribution: Bhutan, China, India, Lao People's Democratic Republic (the), Myanmar, Nepal, Thailand

Coelogyne longirachis Ames

Distribution: Philippines (the)

Coelogyne longpasiaensis J.J.Wood & C.L.Chan

Distribution: Indonesia, Malaysia

Coelogyne lycastoides F.Muell. & Kraenzl.
   Coelogyne whitmeei Schltr.

Distribution: Fiji, New Caledonia (Overseas Territory of France), Samoa, Tonga, Vanuatu

Coelogyne macdonaldii F.Muell. & Kraenzl.
   Coelogyne lamellata Rolfe

Distribution: Fiji, Vanuatu

Coelogyne malintangensis J.J.Sm.

Distribution: Indonesia

Coelogyne malipoensis Z.H.Tsi

Distribution: China, Viet Nam

Coelogyne marmorata Rchb.f.
   Coelogyne zahlbrucknerae Kraenzl.

Distribution: Philippines (the)

Coelogyne marthae S.E.C.Sierra

Distribution: Malaysia

Coelogyne mayeriana Rchb.f.

Distribution: Indonesia, Malaysia

Coelogyne merrillii Ames

Distribution: Philippines (the)

Coelogyne micrantha Lindl.
   Coelogyne clarkei Kraenzl.
   Coelogyne papagena Rchb.f.
   Pleione micrantha (Lindl.) Kuntze

Distribution: India, Myanmar

Coelogyne miniata (Blume) Lindl.
   Chelonanthera miniata Blume
   Coelogyne lauterbachiana Kraenzl.
   Coelogyne simplex Lindl.
   Hologyne lauterbachiana Pfitzer
   Hologyne miniata (Blume) Pfitzer
   Pleione lauterbachiana (Kraenzl.) Kuntze
   Pleione miniata (Blume) Kuntze
   Pleione simplex (Lindl.) Kuntze

Distribution: Indonesia

Coelogyne monilirachis Carr

Distribution: Indonesia, Malaysia

Part II: Coelogyne

Coelogyne monticola J.J.Sm.

Distribution: Indonesia

Coelogyne mooreana Rolfe
Coelogyne psectrantha Gagnep.

Distribution: Viet Nam

Coelogyne mossiae Rolfe

Distribution: India

Coelogyne motleyi Rolfe ex J.J.Wood, D.A.Clayton & C.L.Chan

Distribution: Indonesia, Malaysia

Coelogyne moultonii J.J.Sm.

Distribution: Indonesia, Malaysia

Coelogyne multiflora Schltr.

Distribution: Indonesia

Coelogyne muluensis J.J.Wood

Distribution: Malaysia

Coelogyne naja J.J.Sm.

Distribution: Indonesia, Malaysia

Coelogyne nervosa A.Rich.
Coelogyne corrugata Wight
Pleione corrugata (Wight) Kuntze
Pleione nervosa (A.Rich.) Kuntze

Distribution: India, Myanmar

Coelogyne nitida (Wall. ex D.Don) Lindl.
Coelogyne conferta hort.
Coelogyne corymbosa auct. non Lindl.
Coelogyne ochracea Lindl.
Cymbidium nitidum Wall. ex D.Don non Roxb.
Pleione ochracea (Lindl.) Kuntze

Distribution: Bangladesh, Bhutan, China, India, Lao People's Democratic Republic (the), Myanmar, Nepal, Thailand

Coelogyne obtusifolia Carr

Distribution: Malaysia

Coelogyne occultata Hook.f.
  Pleione occultata (Hook.f.) Kuntze

Distribution: Bhutan, China, India, Myanmar

Coelogyne odoardii Schltr.

Distribution: Brunei Darussalam, Indonesia, Malaysia

Coelogyne odoratissima Lindl.
  Coelogyne angustifolia A.Rich.
  Coelogyne trifida Rchb.f.
  Pleione odoratissima (Lindl.) Kuntze

Distribution: India, Sri Lanka

Coelogyne ovalis Lindl.
  Broughtonia. linearis Wall. ex. Lindl.
  Coelogyne decora Wall. ex. Voigt
  Coelogyne pilosissima Planch.

Distribution: Bhutan, China, India, Myanmar, Nepal, Thailand, Viet Nam

Coelogyne padangensis J.J.Sm. & Schltr.

Distribution: Indonesia

Coelogyne palawanensis Ames

Distribution: Philippines (the)

Coelogyne pallens Ridl.
  Coelogyne ovalis auct. non Lindl.

Distribution: Indonesia, Lao People's Democratic Republic (the), Malaysia, Myanmar,
Thailand, Viet Nam

Coelogyne pandurata Lindl.
  Pleione pandurata (Lindl.) Kuntze

Distribution: Brunei Darussalam, Indonesia, Malaysia

Coelogyne papillosa Ridl. ex Stapf

Distribution: Indonesia, Malaysia

Part II: Coelogyne

Coelogyne parishii Hook.
Pleione parishii (Hook.) Kuntze

Distribution: Myanmar

Coelogyne peltastes Rchb.f.

Distribution: Indonesia, Malaysia

Coelogyne pendula Summerh. ex Parry

Distribution: India

Coelogyne pholidotoides J.J.Sm.

Distribution: Indonesia, Malaysia

Coelogyne picta Schltr.

Distribution: Myanmar

Coelogyne planiscapa Carr

Distribution: Indonesia, Malaysia

Coelogyne plicatissima Ames & C.Schweinf.

Distribution: Indonesia, Malaysia

Coelogyne prasina Ridl.
Coelogyne modesta J.J.Sm.
Coelogyne rhizomatosa J.J.Sm.
Coelogyne vagans Schltr.

Distribution: Indonesia, Malaysia

Coelogyne prolifera Lindl.
Coelogyne flavida Hook.f. ex Lindl.
Pleione flavida (Hook.f. ex Lindl.) Kuntze
Pleione prolifera (Lindl.) Kuntze

Distribution: Bhutan, China, India, Myanmar, Nepal

Coelogyne pulchella Rolfe

Distribution: China, Myanmar

Coelogyne pulverula Teijsm. & Binn.
 Coelogyne dayana Rchb.f.
 Pleione dayana (Rchb.f.) Kuntze

Distribution: Brunei Darussalam, Indonesia, Malaysia, Thailand

Coelogyne punctulata Lindl.
 Coelogyne brevifolia Lindl.
 Coelogyne goweri Rchb.f.
 Coelogyne nitida (Roxb.) Hook.f. non (Wall.ex D.Don) Lindl.
 Coelogyne ocellata Lindl.
 Cymbidium nitidum Roxb.
 Pleione brevifolia (Lindl.) Kuntze
 Pleione goweri (Rchb.f.) Kuntze
 Pleione nitida (Roxb.) Kuntze

Distribution: Bhutan, China, India, Myanmar, Nepal

Coelogyne quadratiloba Gagnep.
 Coelogyne thailandica Seidenf.

Distribution: India, Thailand, Viet Nam

Coelogyne quinquelamellata Ames

Distribution: Philippines (the)

Coelogyne radicosa Ridl.
 Coelogyne carnea Hook.f. non (Blume) Rchb.f.
 Coelogyne radicosus Ridl.
 Coelogyne stipitibulbum Holttum

Distribution: Indonesia, Malaysia, Thailand

Coelogyne radioferens Ames & C.Schweinf.

Distribution: Brunei Darussalam, Indonesia, Malaysia

Coelogyne raizadae S.K.Jain & S.Das
 Coelogyne longipes sensu Hook.f. non Lindl.

Distribution: Bhutan, China, India, Lao People's Democratic Republic (the), Nepal

Coelogyne remediosiae Ames & Quisumb.
Coelogyne remediosae Ames & Quisumb.

Distribution: Philippines (the)

Part II: Coelogyne

Coelogyne renae Gravend. & de Vogel

Distribution: Malaysia

Coelogyne rhabdobulbon Schltr.
Coelogyne pulverula sensu Lamb & C.L.Chan non Teijsm. & Binn.

Distribution: Indonesia, Malaysia

Coelogyne rigida C.S.P.Parish & Rchb.f.
Coelogyne tricarinata Ridl.
Pleione rigida (C.S.P.Parish & Rchb.f.) Kuntze

Distribution: China, India, Malaysia, Myanmar, Thailand, Viet Nam

Coelogyne rigidiformis Ames & C.Schweinf.

Distribution: Indonesia, Malaysia

Coelogyne rochussenii de Vriese
Chelonanthera convallariifolia Blume
Coelogyne convallariifolia (Blume) Zoernig
Coelogyne macrobulbon Hook.f.
Coelogyne plantaginea Lindl.
Coelogyne steffensii Schltr.
Coelogyne stellaris Rchb.f.
Pleione macrobulbon (Hook.f.) Kuntze
Pleione plantaginea (Lindl.) Kuntze
Pleione rochussenii (de Vriese) Kuntze

Distribution: Brunei Darussalam, Indonesia, Malaysia, Philippines (the), Thailand

Coelogyne rumphii Lindl.
Angraecum nervosum Rumph.
Coelogyne psittacina Rchb.f.
Pleione psittacina (Rchb.f.) Kuntze
Pleione rumphii (Lindl.) Kuntze

Distribution: Indonesia

Coelogyne rupicola Carr

Distribution: Indonesia, Malaysia

Coelogyne salmonicolor Rchb.f.
Coelogyne bella Schltr.

Distribution: Indonesia

Coelogyne sanderae Kraenzl. ex O'Brien
Coelogyne annamensis Ridl. non (Lindl. & Rchb.f.) Rolfe
Coelogyne darlacensis Gagnep.
Coelogyne ridleyi Gagnep.

Distribution: China, Myanmar, Viet Nam

Coelogyne sanderiana Rchb.f.
Pleione sanderiana (Rchb.f.) Kuntze

Distribution: Brunei Darussalam, Indonesia, Malaysia

Coelogyne schilleriana Rchb.f. & K.Koch
Pleione schilleriana (Rchb.f.) B.S.Williams

Distribution: Myanmar, Thailand

Coelogyne schultesii S.K.Jain & S.Das
Coelogyne flavida sensu Seidenf. non Hook.f. ex Lindl.
Coelogyne prolifera sensu Lindl.

Distribution: Bhutan, China, India, Lao People's Democratic Republic (the), Myanmar, Nepal Thailand, Viet Nam

Coelogyne septemcostata J.J.Sm.
Coelogyne membranifolia Carr
Coelogyne speciosa sensu Ridl. non (Blume) Lindl. nec Lindl.

Distribution: Brunei Darussalam, Indonesia, Malaysia, Thailand

Coelogyne sparsa Rchb.f.
Pleione sparsa (Rchb.f.) Kuntze

Distribution: Philippines (the)

Coelogyne speciosa (Blume) Lindl.
Chelonanthera speciosa Blume
Pleione speciosa (Blume) Kuntze

Distribution: Indonesia

Coelogyne squamulosa J.J.Sm.

Distribution: Indonesia, Malaysia

Coelogyne steenisii J.J.Sm.

Distribution: Indonesia

Part II: Coelogyne

Coelogyne stenobulbon Schltr.

Distribution: Indonesia

Coelogyne stenochila Hook.f.
  Pleione stenochila (Hook.f.) Kuntze

Distribution: Malaysia

Coelogyne stricta (D.Don) Schltr.
  Coelogyne elata Lindl.
  Cymbidium strictum D.Don
  Pleione elata (Lindl.) Kuntze

Distribution: Bhutan, China, India, Myanmar, Nepal, Viet Nam

Coelogyne suaveolens (Lindl.) Hook.f.
  Coelogyne undulata Wall. ex Pfitzer & Kraenzl. non Rchb.f.
  Pholidota suaveolens Lindl.
  Pleione suaveolens (Lindl.) Kuntze

Distribution: China, India

Coelogyne susanae P.J.Cribb & B.A.Lewis

Distribution: Papua New Guinea, Solomon Islands

Coelogyne swaniana Rolfe
  Coelogyne quadrangularis Ridl.

Distribution: Brunei Darussalam, Indonesia, Malaysia, Philippines (the)

Coelogyne taronensis Hand.-Mazz.

Distribution: China

Coelogyne tenasserimensis Seidenf.

Distribution: Myanmar, Thailand, Viet Nam

Coelogyne tenompokensis Carr

Distribution: Indonesia, Malaysia

Coelogyne tenuis Rolfe
  Coelogyne bihamata J.J.Sm.

Distribution: Indonesia

Coelogyne testacea Lindl.
 Coelogyne sumatrana J.J.Sm.
 Pleione testacea (Lindl.) Kuntze

Distribution: Indonesia, Malaysia, Singapore

Coelogyne tiomanensis M.R.Hend.

Distribution: Malaysia

Coelogyne tomentosa Lindl.
 Coelogyne cymbidioides sensu Ridl. non Rchb.f.
 Coelogyne densiflora Ridl.
 Coelogyne massangeana Rchb.f.
 Pleione massangeana (Rchb.f.) Kuntze
 Pleione tomentosa (Lindl.) Kuntze

Distribution: India, Indonesia, Malaysia, Thailand

Coelogyne tommii Gravend. & P.O'Byrne
 Coelogyne tomiensis P.O'Byrne

Distribution: Indonesia

Coelogyne trilobulata J.J.Sm.

Distribution: Indonesia

Coelogyne trinervis Lindl.
 Coelogyne angustifolia Ridl. non A.Rich. nec Wight
 Coelogyne cinnamomea Lindl.
 Coelogyne pachybulbon Ridl.
 Coelogyne rhodeana Rchb.f.
 Coelogyne rossiana Rchb.f.
 Coelogyne stenophylla Ridl.
 Coelogyne wettsteiniana Schltr.
 Pleione rossiana (Rchb.f.) Kuntze
 Pleione trinervis (Lindl.) Kuntze

Distribution: Cambodia, Indonesia, Lao People's Democratic Republic (the), Malaysia,
Myanmar, Thailand, Viet Nam

Coelogyne triplicatula Rchb.f.

Distribution: Myanmar

Coelogyne triuncialis P.O'Byrne & J.J.Verm.

Distribution: Indonesia

Part II: Coelogyne

Coelogyne tumida J.J.Sm.

Distribution: Indonesia

Coelogyne undatialata J.J.Sm.

Distribution: Indonesia

Coelogyne usitana Roeth & O.Gruss

Distribution: Philippines (the)

Coelogyne ustulata C.S.P.Parish & Rchb.f.
Pleione ustulata (C.S.P.Parish & Rchb.f.) Kuntze

Distribution: Myanmar

Coelogyne vanoverberghii Ames

Distribution: Philippines (the)

Coelogyne veitchii Rolfe

Distribution: Indonesia, Papua New Guinea, Solomon Islands

Coelogyne velutina de Vogel

Distribution: Malaysia, Thailand

Coelogyne venusta Rolfe

Distribution: Brunei Darussalam, Malaysia

Coelogyne vermicularis J.J.Sm.
Chelonistele vermicularis (J.J.Sm.) Kraenzl.

Distribution: Indonesia, Malaysia

Coelogyne verrucosa S.E.C.Sierra

Distribution: Brunei Darussalam, Malaysia

Coelogyne virescens Rolfe

Distribution: Thailand, Viet Nam

Coelogyne viscosa Rchb.f.
Coelogyne graminifolia C.S.P.Parish & Rchb.f.
Pleione graminifolia (C.S.P.Parish & Rchb.f.) Kuntze
Pleione viscosa (Rchb.f.) Kuntze

Distribution: Bangladesh, China, India, Lao People's Democratic Republic (the), Malaysia, Myanmar, Thailand, Viet Nam

Coelogyne xyrekes Ridl.
Coelogyne xanthoglossa Ridl.

Distribution: Indonesia, Malaysia, Thailand

Coelogyne yiii Schuit. & de Vogel

Distribution: Malaysia

Coelogyne zeylanica Hook.f.

Distribution: India, Sri Lanka

Coelogyne zhenkangensis S.C.Chen & K.Y.Lang

Distribution: China

Coelogyne zurowetzii Carr

Distribution: Indonesia, Malaysia

COMPARETTIA BINOMIALS IN CURRENT USE

COMPARETTIA BINOMES ACTUELLEMENT EN USAGE

COMPARETTIA BINOMIALES UTILIZADOS NORMALMENTE

Comparettia coccinea Lindl.
  Comparettia peruviana Schltr.

Distribution: Bolivia, Brazil, Colombia, Peru

Comparettia falcata Poepp. & Endl.
  Comparettia cryptocera C.Morren
  Comparettia erecta Schltr.
  Comparettia pulchella Schltr.
  Comparettia rosea Lindl.
  Comparettia venezuelana Schltr.

Distribution: Belize, Brazil, Colombia, Costa Rica, Cuba, Guatemala, Honduras, Mexico, Peru, Puerto Rico (Dependant Territorty of the United States of America), United States of America

Comparettia ignea P.Ortíz

Distribution: Colombia, Peru

Comparettia macroplectron Rchb.f. & Triana
  Comparettia splendens Schltr.

Distribution: Colombia

Comparettia × maloi I.Bock

Distribution: Ecuador

Comparettia speciosa Rchb.f.

Distribution: Ecuador

MASDEVALLIA BINOMIALS IN CURRENT USE

MASDEVALLIA BINOMES ACTUELLEMENT EN USAGE

MASDEVALLIA BINOMIALES UTILIZADOS NORMALMENTE

Masdevallia abbreviata Rchb.f.

Distribution: Ecuador, Peru

Masdevallia acaroi Luer & Hirtz

Distribution: Ecuador

Masdevallia acrochordonia Rchb.f.

Distribution: Ecuador

Masdevallia adamsii Luer

Distribution: Belize

Masdevallia adrianae Luer

Distribution: Ecuador

Masdevallia aenigma Luer & R.Escobar

Distribution: Colombia

Masdevallia agaster Luer

Distribution: Ecuador

Masdevallia aguirrei Luer & R.Escobar

Distribution: Colombia

Masdevallia akemiana Königer & Sijm

Distribution: Colombia

Masdevallia albella Luer & Teague

Distribution: Ecuador, Peru

Part II: Masdevallia

Masdevallia alexandri Luer

Distribution: Ecuador

Masdevallia alismifolia Kraenzl.

Distribution: Colombia

Masdevallia × alvaroi Luer & R.Escobar

Distribution: Colombia

Masdevallia amabilis Rchb.f.& Warsz.
  Masdevallia flammula H.Mohr & Braas
  Masdevallia purpurina Schltr.

Distribution: Peru

Masdevallia amaluzae Luer & Malo

Distribution: Ecuador

Masdevallia amanda Rchb.f.& Warsz.
  Masdevallia calopterocarpa Rchb.f.
  Masdevallia gustavii Rchb.f.
  Masdevallia haematocantha Lindl sensu Woolward nom.nud.
  Masdevallia oligantha Schltr.
  Masdevallia remotiflora Kraenzl.

Distribution: Colombia, Ecuador, Venezuela

Masdevallia ametroglossa Luer & Hirtz

Distribution: Ecuador

Masdevallia amoena Luer

Distribution: Ecuador

Masdevallia amplexa Luer

Distribution: Peru

Masdevallia ampullacea Luer & Andreetta

Distribution: Ecuador

Masdevallia anceps Luer & Hirtz

Distribution: Ecuador

Masdevallia andreettaeana Luer
Masdevallia andreettana Luer

Distribution: Ecuador, Peru

Masdevallia anemone Luer

Distribution: Ecuador

Masdevallia anfracta Königer & J.Portilla

Distribution: Ecuador

Masdevallia angulata Rchb.f.
Masdevallia burfordiensis hort. ex O'Brien

Distribution: Colombia, Ecuador

Masdevallia angulifera Rchb.f.ex Kraenzl.
Masdevallia olivacea Kraenzl.

Distribution: Colombia

Masdevallia anisomorpha Garay

Distribution: Colombia

Masdevallia anomala Luer & Sijm

Distribution: Peru

Masdevallia antonii Königer

Distribution: Peru

Masdevallia aphanes Königer

Distribution: Ecuador, Peru

Masdevallia apparitio Luer & R.Escobar

Distribution: Colombia

Part II: Masdevallia

Masdevallia aptera Luer & L.O'Shaughn.

Distribution: Ecuador

Masdevallia arangoi Luer & R.Escobar

Distribution: Colombia

Masdevallia ariasii Luer

Distribution: Ecuador, Peru

Masdevallia arminii Linden & Rchb.f.

Distribution: Colombia

Masdevallia assurgens Luer & R.Escobar

Distribution: Colombia

Masdevallia asterotricha Königer

Distribution: Peru

Masdevallia atahualpa Luer

Distribution: Peru

Masdevallia attenuata Rchb.f.
   Masdevallia fonsecae Königer

Distribution: Costa Rica, Ecuador, Panama

Masdevallia audax Königer

Distribution: Peru

Masdevallia aurea Luer
   Masdevallia santiagoörum Königer

Distribution: Ecuador

Masdevallia aurorae Luer & M.W.Chase

Distribution: Peru

Masdevallia ayabacana Luer

Distribution: Peru

Masdevallia bangii Schltr.
 Masdevallia trioön Sweet
 Physosiphon bangii (Schltr.) Garay

Distribution: Bolivia, Ecuador, Peru

Masdevallia barlaeana Rchb.f.

Distribution: Peru

Masdevallia barrowii Luer

Distribution: Ecuador

Masdevallia belua Königer & D'Aless.

Distribution: Ecuador

Masdevallia bennettii Luer

Distribution: Peru

Masdevallia berthae Luer & Andreetta

Distribution: Ecuador

Masdevallia bicolor Poepp & Endl.
 Masdevallia atropurpurea Rchb.f.(sphalm.)
 Masdevallia aureorosea Weberb.
 Masdevallia auropurpurea Rchb.f.& Warsz.
 Masdevallia biflora E.Morren
 Masdevallia herzogii Schltr.
 Masdevallia peruviana Rolfe
 Masdevallia subumbellata Kraenzl.
 Masdevallia xanthura Schltr.

Distribution: Bolivia, Colombia, Ecuador, Peru, Venezuela

Masdevallia bicornis Luer
 Portillia popowiana Königer & J.Portilla

Distribution: Ecuador

Part II: Masdevallia

Masdevallia boliviensis Schltr.
Masdevallia leucophaea Luer & R.Vasquez

Distribution: Bolivia

Masdevallia bonplandii Rchb.f.
Masdevallia endotrachys Kraenzl.
Masdevallia sulphurea F.Lehm & Kraenzl.
Masdevallia uniflora Kunth non Ruiz & Pav.

Distribution: Colombia, Ecuador, Peru

Masdevallia bottae Luer & Andreetta

Distribution: Ecuador

Masdevallia bourdetteana Luer

Distribution: Ecuador

Masdevallia brachyantha Schltr.

Distribution: Bolivia

Masdevallia brachyura F.Lehm & Kraenzl.

Distribution: Ecuador

Masdevallia brenneri Luer

Distribution: Ecuador

Masdevallia brockmuelleri Luer

Distribution: Colombia

Masdevallia bryophila Luer
Masdevallia trechsliniana Königer & J.Meza

Distribution: Peru

Masdevallia buccinator Rchb.f.& Warsz.

Distribution: Colombia

Masdevallia bucculenta Luer & Hirtz

Distribution: Ecuador

Masdevallia bulbophyllopsis Kraenzl.
  Masdevallia invenusta Luer

Distribution: Ecuador

Masdevallia burianii Luer & Dalstrom

Distribution: Bolivia

Masdevallia cacodes Luer & R.Escobar

Distribution: Colombia

Masdevallia caesia Roezl
  Masdevallia deorsum Rolfe
  Masdevallia metallica F.Lehm & Kraenzl.

Distribution: Colombia

Masdevallia calagrasalis Luer

Distribution: Ecuador

Masdevallia calocalix Luer

Distribution: Ecuador

Masdevallia caloptera Rchb.f.
  Masdevallia biflora Regel non E.Morren

Distribution: Peru

Masdevallia calosiphon Luer

Distribution: Peru

Masdevallia calura Rchb.f.

Distribution: Costa Rica

Part II: Masdevallia

Masdevallia campyloglossa Rchb.f.
Masdevallia campyloglossa Rolfe non Rchb.f.
Masdevallia dermatantha Kraenzl.
Masdevallia fertilis Kraenzl.
Masdevallia heterotepala Rchb.f.
Masdevallia ortgiesiana hort. ex Rolfe
Masdevallia sarcophylla Kraenzl.

Distribution: Colombia, Ecuador, Peru

Masdevallia cardiantha Königer

Distribution: Peru

Masdevallia carmenensis Luer & Malo

Distribution: Ecuador

Masdevallia carnosa Königer

Distribution: Peru

Masdevallia carpishica Luer & Cloes

Distribution: Peru

Masdevallia carruthersiana F.Lehm & Kraenzl.
Masdevallia margaretae Luer
Masdevallia torulosa Königer & J.Portilla

Distribution: Ecuador

Masdevallia castor Luer & Cloes

Distribution: Peru

Masdevallia catapheres Königer

Distribution: Peru

Masdevallia caudata Lindl.
Masdevallia cucutillensis Kraenzl.
Masdevallia shuttleworthii Rchb.f.
Masdevallia tricolor Rchb.f.

Distribution: Colombia, Venezuela

Masdevallia caudivolvula Kraenzl.

Distribution: Colombia

Masdevallia cerastes Luer & R.Escobar

Distribution: Colombia

Masdevallia chaetostoma Luer

Distribution: Ecuador

Masdevallia chaparensis T.Hashim.
  Masdevallia hajekii Luer

Distribution: Bolivia

Masdevallia chasei Luer

Distribution: Costa Rica

Masdevallia chimboënsis Kraenzl.

Distribution: Colombia, Ecuador

Masdevallia chontalensis Rchb.f.
  Masdevallia diantha Schltr.

Distribution: Costa Rica, Guatemala, Nicaragua, Panama,Venezuela

Masdevallia chuspipatae Luer & Teague

Distribution: Bolivia

Masdevallia cinnamomea Rchb.f.

Distribution: Peru

Masdevallia citrinella Luer & Malo

Distribution: Ecuador

Masdevallia civilis Rchb.f.& Warsz.
  Masdevallia aequiloba Regel

Distribution: Peru

Part II: Masdevallia

Masdevallia clandestina Luer & R.Escobar

Distribution: Colombia, Venezuela

Masdevallia cleistogama Luer

Distribution: Peru

Masdevallia cloesii Luer

Distribution: Peru

Masdevallia cocapatae Luer, Teague & R.Vasquez

Distribution: Bolivia

Masdevallia coccinea Linden ex Lindl.
  Masdevallia approviata hort. ex Woolward
  Masdevallia armeniaca B.S.Williams
  Masdevallia atrosanguinea B.S.Williams
  Masdevallia cinnabarina Linden ex sic.
  Masdevallia coerulescens B.S.Williams
  Masdevallia denisonii Dombrain
  Masdevallia harryana Rchb.f.
  Masdevallia laeta Rchb.f.
  Masdevallia lateritia hort. nom.nud.
  Masdevallia lindenii André
  Masdevallia longiflora Cogn.
  Masdevallia militaris Rchb.f.& Warsz.
  Masdevallia miniata B.S.Williams
  Masdevallia musaica E Morr.
  Masdevallia tricolor hort. non Rchb.f.
  Masdevallia versicolor hort.

Distribution: Colombia

Masdevallia collantesii D.E.Benn & Christenson

Distribution: Peru

Masdevallia collina L.O.Williams

Distribution: Panama

Masdevallia colossus Luer

Distribution: Ecuador, Peru

Masdevallia concinna Königer

Distribution: Peru

Masdevallia condorensis Luer & Hirtz

Distribution: Ecuador

Masdevallia constricta Poepp & Endl.
  Masdevallia urosalpinx Luer

Distribution: Ecuador, Peru

Masdevallia corazonica Schltr.
  Masdevallia sphenopetala Kraenzl.

Distribution: Ecuador

Masdevallia cordeliana Luer

Distribution: Peru

Masdevallia corderoana F.Lehm & Kraenzl.

Distribution: Ecuador

Masdevallia coriacea Lindl.
  Masdevallia bogotensis hort. ex Gentil
  Masdevallia bruckmuelleri Linden & André
  Masdevallia ochracea Burb nom.nud.

Distribution: Colombia, Ecuador, Peru

Masdevallia corniculata Rchb.f.
  Masdevallia calyptrata Kraenzl.
  Masdevallia eclyptrata Kraenzl (sphalm)
  Masdevallia inflata Rchb.f.

Distribution: Colombia, Ecuador

Masdevallia cosmia Königer

Distribution: Peru

Masdevallia cranion Luer

Distribution: Peru

Part II: Masdevallia

Masdevallia crassicaudis Luer & J.Portilla
Masdevallia blanda Königer & J.Portilla
Masdevallia crassicaulis Luer & J.Portilla (sphalm.)

Distribution: Ecuador

Masdevallia crescenticola F.Lehm & Kraenzl.

Distribution: Colombia, Ecuador

Masdevallia cretata Luer

Distribution: Ecuador

Masdevallia cucullata Lindl.

Distribution: Colombia, Ecuador

Masdevallia cuprea Lindl.
Masdevallia cayennensis Rchb.f.
Masdevallia hepatica Luer
Masdevallia manningii Königer

Distribution: Brazil, Ecuador, French Guiana (Overseas Department of France), Peru, Suriname

Masdevallia cupularis Rchb.f.
Masdevallia odontochila Schltr.
Masdevallia reflexa Schltr.

Distribution: Costa Rica

Masdevallia curtipes Barb.Rodr.

Distribution: Brazil

Masdevallia cyclotega Königer
Masdevallia mijahuangae D.E.Benn.

Distribution: Peru

Masdevallia cylix Luer & Malo

Distribution: Ecuador

Masdevallia dalessandroi Luer
 Masdevallia velox Königer

Distribution: Ecuador

Masdevallia dalstroemii Luer

Distribution: Ecuador

Masdevallia datura Luer & R.Vasquez

Distribution: Bolivia

Masdevallia davisii Rchb.f.

Distribution: Peru

Masdevallia deceptrix Luer & Würstle

Distribution: Venezuela

Masdevallia decumana Königer

Distribution: Ecuador, Peru

Masdevallia deformis Kraenzl.
 Masdevallia exaltata Luer

Distribution: Ecuador

Masdevallia delhierroi Luer & Hirtz

Distribution: Ecuador

Masdevallia delphina Luer

Distribution: Ecuador

Masdevallia demissa Rchb.f.

Distribution: Costa Rica

Masdevallia deniseana Luer & J.Portilla

Distribution: Ecuador

Part II: Masdevallia

Masdevallia densiflora Schltr.

Distribution: Colombia

Masdevallia descendens Luer & Andreetta

Distribution: Ecuador

Masdevallia dimorphotricha Luer & Hirtz

Distribution: Ecuador

Masdevallia discoidea Luer & Würstle

Distribution: Brazil

Masdevallia discolor Luer & R.Escobar

Distribution: Colombia

Masdevallia don-quijote Luer & Andreetta

Distribution: Ecuador

Masdevallia dorisiae Luer

Distribution: Ecuador

Masdevallia draconis Luer & Andreetta

Distribution: Ecuador

Masdevallia dreisei Luer

Distribution: Ecuador

Masdevallia dryada Luer & R.Escobar

Distribution: Colombia

Masdevallia dudleyi Luer

Distribution: Peru

Masdevallia dunstervillei Luer

Distribution: Venezuela

Masdevallia dura Luer

Distribution: Ecuador

Masdevallia dynastes Luer

Distribution: Ecuador

Masdevallia eburnea Luer & Maduro

Distribution: Panama

Masdevallia echo Luer

Distribution: Peru

Masdevallia ejiriana Luer & J.Portilla

Distribution: Ecuador

Masdevallia elachys Luer

Distribution: Bolivia

Masdevallia elegans Luer & R.Escobar

Distribution: Peru

Masdevallia elephanticeps Rchb.f.& Warsz.

Distribution: Colombia

Masdevallia empusa Luer
  Masdevallia jimenezii Königer

Distribution: Ecuador, Peru

Masdevallia encephala Luer & R.Escobar

Distribution: Colombia

Part II: Masdevallia

Masdevallia ensata Rchb.f.
Masdevallia schildhaueri Königer
Masdevallia xiphium Rchb.f.ex Kraenzl.

Distribution: Venezuela

Masdevallia epallax Königer
Masdevallia enallax Königer

Distribution: Costa Rica

Masdevallia ephelota Luer & Cloes

Distribution: Peru

Masdevallia estradae Rchb.f.

Distribution: Colombia

Masdevallia eucharis Luer

Distribution: Ecuador

Masdevallia eumeces Luer

Distribution: Peru

Masdevallia eumeliae Luer

Distribution: Peru

Masdevallia eurynogaster Luer & Andreetta

Distribution: Ecuador

Masdevallia excelsior Luer & Andreetta

Distribution: Ecuador

Masdevallia expansa Rchb.f.

Distribution: Colombia

Masdevallia expers Luer & Andreetta

Distribution: Ecuador

Masdevallia exquisita Luer & Hirtz

Distribution: Bolivia

Masdevallia falcago Rchb.f.
Masdevallia trionyx Kraenzl.

Distribution: Colombia

Masdevallia fasciata Rchb.f.
Masdevallia bathyschista Schltr.
Masdevallia palmensis Kraenzl.
Masdevallia restrepioidea Kraenzl.
Masdevallia trinemoides Kraenzl.
Rodrigoa fasciata (Rchb.f.) Braas

Distribution: Colombia

Masdevallia figueroae Luer

Distribution: Ecuador, Peru

Masdevallia filaria Luer & R.Escobar

Distribution: Colombia, Ecuador

Masdevallia flaveola Rchb.f.

Distribution: Costa Rica, Panama

Masdevallia floribunda Lindl.
Masdevallia burzlaffiana Königer
Masdevallia floribunda Ames non Lindl.
Masdevallia galeottiana A.Rich & Galeotti
Masdevallia lindeniana A.Rich & Galeotti
Masdevallia myriostigma Morren
Masdevallia tuerckheimii (Ames) Luer

Distribution: Belize, Colombia, Costa Rica, Guatemala, Honduras, Mexico, Colombia

Masdevallia foetens Luer & R.Escobar

Distribution: Colombia

Masdevallia formosa Luer & Cloes

Distribution: Peru

Part II: Masdevallia

Masdevallia fosterae Luer

Distribution: Unknown

Masdevallia fractiflexa F.Lehm & Kraenzl.

Distribution: Ecuador

Masdevallia fragrans Woolward

Distribution: Colombia

Masdevallia frilehmannii Luer & R.Vasquez

Distribution: Bolivia

Masdevallia fuchsii Luer
  Masdevallia saulii Königer

Distribution: Peru

Masdevallia fulvescens Rolfe

Distribution: Costa Rica

Masdevallia garciae Luer

Distribution: Venezuela

Masdevallia gargantua Rchb.f.

Distribution: Colombia

Masdevallia geminiflora P.Ortíz

Distribution: Colombia, Ecuador

Masdevallia gilbertoi Luer & R.Escobar

Distribution: Colombia

Masdevallia glandulosa Königer

Distribution: Ecuador, Peru

Masdevallia glomerosa Luer & Andreetta

Distribution: Ecuador

Masdevallia gloriae Luer & Maduro

Distribution: Panama

Masdevallia gnoma Sweet

Distribution: Ecuador

Masdevallia goliath Luer & Andreetta

Distribution: Ecuador, Peru

Masdevallia graminea Luer

Distribution: Ecuador

Masdevallia guayanensis Lindl ex Benth.
  Masdevallia manarana Carnevali & Ramírez

Distribution: Guyana, Venezuela

Masdevallia guerrieroi Luer & Andreetta

Distribution: Ecuador

Masdevallia gutierrezii Luer

Distribution: Bolivia

Masdevallia guttulata Rchb.f.
  Masdevallia guttulata Rolfe non Rchb.f.
  Masdevallia lawrencei Kraenzl.

Distribution: Ecuador

Masdevallia harlequina Luer
  Masdevallia shiraishii Königer & M.Arias

Distribution: Peru

Masdevallia hartmanii Luer

Distribution: Ecuador

Part II: Masdevallia

Masdevallia heideri Königer

Distribution: Bolivia

Masdevallia helenae Luer

Distribution: Bolivia

Masdevallia helgae Königer & J.Portilla

Distribution: Ecuador

Masdevallia henniae Luer & Dalstrom

Distribution: Ecuador

Masdevallia hercules Luer & Andreetta

Distribution: Colombia, Ecuador

Masdevallia herradurae F.Lehm & Kraenzl.
Masdevallia frontinoënsis Kraenzl.
Masdevallia tenuipes Schltr.

Distribution: Colombia

Masdevallia heteroptera Rchb.f.
Masdevallia fissa Kraenzl.
Masdevallia heteromorpha Rchb.f.
Rodrigoa heteroptera (Rchb.f.) Braas

Distribution: Colombia

Masdevallia hians Linden & Rchb.f.
Masdevallia copiosa Kraenzl.
Masdevallia nivalis Rchb.f.ex Kraenzl nom.nud.

Distribution: Colombia

Masdevallia hieroglyphica Rchb.f.

Distribution: Colombia

Masdevallia hirtzii Luer & Andreetta

Distribution: Ecuador

Masdevallia hortensis Luer & R.Escobar

Distribution: Colombia

Masdevallia hubeinii Luer & Würstle

Distribution: Colombia

Masdevallia hydrae Luer

Distribution: Ecuador

Masdevallia hylodes Luer & R.Escobar

Distribution: Colombia

Masdevallia hymenantha Rchb.f.

Distribution: Peru

Masdevallia hystrix Luer & Hirtz

Distribution: Ecuador

Masdevallia icterina Königer

Distribution: Peru

Masdevallia idae Luer & Arias

Distribution: Peru

Masdevallia ignea Rchb.f.
Masdevallia boddaertii Linden ex André
Masdevallia coccinea Regel non Linden ex Lindl.

Distribution: Colombia

Masdevallia immensa Luer

Distribution: Peru

Masdevallia impostor Luer & R.Escobar

Distribution: Colombia, Ecuador, Venezuela

Part II: Masdevallia

Masdevallia indecora Luer & R.Escobar

Distribution: Colombia

Masdevallia infracta Lindl.
Masdevallia albida Lem.
Masdevallia aristata Barb.Rodr.
Masdevallia aurantiaca Lindl.
Masdevallia forgetiana Kraenzl.
Masdevallia longicaudata Lem.
Masdevallia tridentata Lindl.
Masdevallia triquetra Scheidw.

Distribution: Bolivia, Brazil

Masdevallia ingridiana Luer & J.Portilla

Distribution: Ecuador

Masdevallia instar Luer & Andreetta

Distribution: Ecuador, Peru

Masdevallia ionocharis Rchb.f.

Distribution: Peru

Masdevallia irapana H.R.Sweet

Distribution: Venezuela

Masdevallia iris Luer & R.Escobar

Distribution: Venezuela

Masdevallia ishikoi Luer

Distribution: Bolivia

Masdevallia isos Luer

Distribution: Bolivia

Masdevallia jarae Luer

Distribution: Peru

Masdevallia josei Luer

Distribution: Ecuador

Masdevallia juan-albertoi Luer & M.Arias

Distribution: Peru

Masdevallia karineae Nauray ex Luer

Distribution: Peru

Masdevallia klabochiorum Rchb.f.
  Masdevallia aops Luer & Malo
  Masdevallia exilipes Schltr.
  Masdevallia paisbambae Kraenzl.

Distribution: Colombia, Ecuador, Peru

Masdevallia kuhniorum Luer

Distribution: Peru

Masdevallia kyphonantha H.R.Sweet
  Masdevallia pseudominuta H.R.Sweet

Distribution: Venezuela

Masdevallia laevis Lindl.
  Masdevallia affinis Lindl.
  Masdevallia chlorotica Kraenzl.
  Masdevallia confusa Kraenzl.
  Masdevallia gomeziana F.Lehm & Kraenzl.
  Masdevallia lepida Rchb.f.
  Masdevallia maculigera Schltr.
  Masdevallia pantherina F.Lehm & Kraenzl.
  Masdevallia petiolaris Schltr.

Distribution: Colombia, Ecuador

Masdevallia lamia Luer & Hirtz

Distribution: Ecuador

Masdevallia lamprotyria Königer

Distribution: Ecuador, Peru

Part II: Masdevallia

Masdevallia lankesterana Luer

Distribution: Costa Rica

Masdevallia lansbergii Rchb.f.
Physosiphon lansbergii (Rchb.f.) L.O.Williams

Distribution: French Guiana (Overseas Department of France), Venezuela

Masdevallia lappifera Luer & Hirtz

Distribution: Ecuador

Masdevallia lata Rchb.f.
Masdevallia borucana P.H.Allen

Distribution: Costa Rica, Panama

Masdevallia laucheana Kraenzl.

Distribution: Costa Rica

Masdevallia leathersii Luer

Distribution: Ecuador

Masdevallia lehmannii Rchb.f.

Distribution: Ecuador

Masdevallia lenae Luer & Hirtz
Masdevallia gracilior Königer & J.Portilla

Distribution: Ecuador

Masdevallia leonardoi Luer

Distribution: Ecuador

Masdevallia leonii D.E.Benn & Christenson

Distribution: Peru

Masdevallia leontoglossa Rchb.f.

Distribution: Colombia

Masdevallia leptoura Luer

Distribution: Colombia, Ecuador, Peru

Masdevallia leucantha F.Lehm & Kraenzl.
  Masdevallia chiquindensis Kraenzl.

Distribution: Ecuador

Masdevallia lewisii Luer & R.Vasquez

Distribution: Bolivia

Masdevallia × **ligiae** Luer & R.Escobar

Distribution: Colombia

Masdevallia lilacina Königer

Distribution: Ecuador, Peru

Masdevallia lilianae Luer
  Masdevallia melina Königer & J.Meza

Distribution: Peru

Masdevallia limax Luer

Distribution: Ecuador

Masdevallia lineolata Königer

Distribution: Peru

Masdevallia lintricula Königer

Distribution: Ecuador, Peru

Masdevallia livingstoneana Roezl
  Masdevallia panamensis (Schltr.) Ames
  Scaphosepalum panamense Schltr.

Distribution: Colombia, Costa Rica, Panama

Part II: Masdevallia

Masdevallia loui Luer & Dalstrom

Distribution: Ecuador

Masdevallia lucernula Königer

Distribution: Peru

Masdevallia ludibunda Rchb.f.

Distribution: Colombia

Masdevallia ludibundella Luer & R.Escobar

Distribution: Colombia

Masdevallia luziae-mariae Luer & R.Vasquez

Distribution: Bolivia

Masdevallia lychniphora Königer

Distribution: Peru

Masdevallia lynniana Luer

Distribution: Ecuador

Masdevallia macrogenia (Arango) Luer & R.Escobar

Distribution: Colombia

Masdevallia macroglossa Rchb.f.
 Masdevallia gerlachii Königer
 Masdevallia porcelliceps Rchb.f.

Distribution: Colombia, Venezuela

Masdevallia macropus F.Lehm & Kraenzl.

Distribution: Ecuador

Masdevallia macrura Rchb.f.

Distribution: Colombia

Masdevallia maculata Klotzsch & H.Karst.

Distribution: Venezuela

Masdevallia maduroi Luer

Distribution: Panama

Masdevallia mallii Luer

Distribution: Ecuador

Masdevallia maloi Luer

Distribution: Ecuador

Masdevallia manchinazae Luer & Andreetta

Distribution: Ecuador

Masdevallia mandarina (Luer & R.Escobar) Luer

Distribution: Colombia, Ecuador

Masdevallia manoloi Luer & M.Arias

Distribution: Peru

Masdevallia manta Königer & Sijm

Distribution: Ecuador

Masdevallia marginella Rchb.f.
 Masdevallia costaricensis Rolfe

Distribution: Costa Rica

Masdevallia marizae Luer & Rolando

Distribution: Peru

Masdevallia marthae Luer & R.Escobar

Distribution: Colombia

Part II: Masdevallia

Masdevallia martineae Luer

Distribution: Bolivia

Masdevallia martiniana Luer

Distribution: Ecuador

Masdevallia mascarata Luer & R.Vasquez

Distribution: Bolivia

Masdevallia mastodon Rchb.f.

Distribution: Colombia

Masdevallia mataxa Königer & H.Mend.

Distribution: Ecuador

Masdevallia maxilimax (Luer) Luer

Distribution: Ecuador

Masdevallia mayaycu Luer & Andreetta

Distribution: Ecuador

Masdevallia medinae Luer & J.Portilla

Distribution: Ecuador

Masdevallia medusa Luer & R.Escobar

Distribution: Colombia

Masdevallia mejiana Garay

Distribution: Colombia

Masdevallia melanoglossa Luer

Distribution: Ecuador

Masdevallia melanopus Rchb.f.

Distribution: Peru

Masdevallia melanoxantha Linden & Rchb.f.
  Masdevallia asperrima Kraenzl.

Distribution: Colombia, Venezuela

Masdevallia meleagris Lindl.
  Rodrigoa meleagris (Lindl.) Braas

Distribution: Colombia

Masdevallia menatoi Luer & R.Vasquez
  Masdevallia foeda Luer & R.Vasquez

Distribution: Bolivia

Masdevallia mendozae Luer

Distribution: Ecuador

Masdevallia mentosa Luer
  Masdevallia glossacles Luer

Distribution: Ecuador

Masdevallia merinoi Luer & J.Portilla

Distribution: Ecuador

Masdevallia mezae Luer
  Masdevallia rauhii Senghas & Braas

Distribution: Peru

Masdevallia microptera Luer & Würstle

Distribution: Peru

Masdevallia microsiphon Luer

Distribution: Ecuador

Part II: Masdevallia

Masdevallia midas Luer

Distribution: Ecuador

Masdevallia milagroi Luer & Hirtz

Distribution: Ecuador

Masdevallia minuta Lindl.
  Masdevallia surinamensis Focke

Distribution: Bolivia, Brazil, Colombia, Ecuador, French Guiana (Overseas Department of France), Guyana, Peru, Suriname,Venezuela

Masdevallia misasii Braas
  Masdevallia reflexa Misas non Schltr.

Distribution: Colombia

Masdevallia molossoides Kraenzl.
  Masdevallia anura Kraenzl.
  Masdevallia rhopalura Schltr.

Distribution: Costa Rica, Nicaragua, Panama

Masdevallia molossus Rchb.f.
  Masdevallia antioquiensis F.Lehm & Kraenzl.
  Masdevallia schmidtchenii Kraenzl.

Distribution: Colombia

Masdevallia × monicana Luer

Distribution: Ecuador

Masdevallia monogona Königer

Distribution: Peru

Masdevallia mooreana Rchb.f.
  Masdevallia atroviolacea Kraenzl.
  Masdevallia sororcula Rchb.f.

Distribution: Colombia

Masdevallia morochoi Luer & Andreetta

Distribution: Ecuador

124

Masdevallia murex Luer

Distribution: Ecuador

Masdevallia mutica Luer & R.Escobar

Distribution: Colombia

Masdevallia × mystica Luer

Distribution: Colombia

Masdevallia naranjapatae Luer

Distribution: Ecuador

Masdevallia navicularis Garay & Dunst.
  Masdevallia scapha Braas

Distribution: Venezuela

Masdevallia nebulina Luer

Distribution: Bolivia

Masdevallia newmaniana Luer & Teague

Distribution: Ecuador

Masdevallia nicaraguae Luer

Distribution: Nicaragua

Masdevallia nidifica Rchb.f.
  Masdevallia cyathogastra Schltr.
  Masdevallia tenuicaudata Schltr.

Distribution: Colombia, Costa Rica, Ecuador, Nicaragua, Panama

Masdevallia niesseniae Luer

Distribution: Colombia

Masdevallia nigricans Königer & Sijm

Distribution: Ecuador

Part II: Masdevallia

Masdevallia nikoleana Luer & J.Portilla

Distribution: Ecuador, Peru

Masdevallia nitens Luer

Distribution: Bolivia

Masdevallia nivea (Luer & R.Escobar) Luer & R.Escobar

Distribution: Colombia

Masdevallia norae Luer

Distribution: Brazil, Colombia, Venezuela

Masdevallia norops Luer & Andreetta

Distribution: Ecuador, Peru

Masdevallia notosibirica F.Maek & T.Hashim.

Distribution: Bolivia

Masdevallia obscurans (Luer) Luer

Distribution: Brazil

Masdevallia odontocera Luer & R.Escobar

Distribution: Colombia

Masdevallia odontopetala Luer

Distribution: Ecuador

Masdevallia omorenoi Luer & R.Vasquez

Distribution: Bolivia

Masdevallia ophioglossa Rchb.f.
Masdevallia grossa Luer

Distribution: Ecuador

Masdevallia oreas Luer & R.Vasquez

Distribution: Bolivia

Masdevallia ortalis Luer

Distribution: Peru

Masdevallia os-draconis Luer & R.Escobar

Distribution: Colombia

Masdevallia os-viperae Luer & Andreetta
  Masdevallia rana-aurea Luer

Distribution: Ecuador, Peru

Masdevallia oscarii Luer & R.Escobar

Distribution: Colombia

Masdevallia oscitans (Luer) Luer

Distribution: Brazil

Masdevallia ostaurina Luer & V.N.M.Rao

Distribution: Panama

Masdevallia ova-avis Luer

Distribution: Ecuador

Masdevallia oxapampaensis D.E.Benn & Christenson

Distribution: Peru

Masdevallia pachyantha Rchb.f.
  Masdevallia hoppii Schltr.

Distribution: Colombia

Masdevallia pachygyne Kraenzl.

Distribution: Colombia

Part II: Masdevallia

Masdevallia pachysepala (Rchb.f.) Luer

Distribution: Colombia, Venezuela

Masdevallia pachyura Rchb.f.
Masdevallia aureodactyla Luer

Distribution: Ecuador

Masdevallia paivaëana Rchb.f.
Masdevallia aspera Rchb.f.ex Kraenzl.

Distribution: Bolivia

Masdevallia pandurilabia C.Schweinf.

Distribution: Peru

Masdevallia panguiënsis Luer & Andreetta

Distribution: Ecuador

Masdevallia pantomima Luer & Hirtz

Distribution: Ecuador

Masdevallia papillosa Luer

Distribution: Ecuador

Masdevallia paquishae Luer & Hirtz
Masdevallia trivenia Königer

Distribution: Ecuador, Peru

Masdevallia pardina Rchb.f.
Masdevallia aequatorialis F.Lehm & Kraenzl.

Distribution: Colombia, Ecuador

Masdevallia parvula Schltr.
Masdevallia diversifolia Kraenzl.
Rodrigoa diversifolia (Kraenzl.) Braas

Distribution: Bolivia, Colombia, Ecuador, Peru

Masdevallia pastinata Luer

Distribution: Colombia

Masdevallia patchicutzae Luer & Hirtz

Distribution: Ecuador

Masdevallia patriciana Luer

Distribution: Ecuador

Masdevallia patula Luer & Malo

Distribution: Ecuador

Masdevallia peristeria Rchb.f.
  Masdevallia ellipes Rchb.f.
  Masdevallia haematosticta Rchb.f.

Distribution: Colombia, Ecuador

Masdevallia pernix Königer

Distribution: Peru

Masdevallia persicina Luer

Distribution: Ecuador

Masdevallia pescadoënsis Luer & R.Escobar

Distribution: Colombia

Masdevallia phacopsis Luer & Dalstrom

Distribution: Bolivia

Masdevallia phasmatodes Königer

Distribution: Peru

Masdevallia phlogina Luer

Distribution: Peru

Part II: Masdevallia

Masdevallia phoenix Luer

Distribution: Peru

Masdevallia picea Luer
  Masdevallia rufolutea Lindl.

Distribution: Peru

Masdevallia picta Luer

Distribution: Ecuador, Peru

Masdevallia picturata Rchb.f.
  Masdevallia cryptocopis Rchb.f.ex Kraenzl.
  Masdevallia meleagris Lindl sensu Rchb.f.non Lindl.
  Masdevallia ocanensis Kraenzl.
  Rodrigoa cryptocopis (Kraenzl.) Braas

Distribution: Bolivia, Colombia, Costa Rica, Ecuador, Guyana, Panama, Peru, Venezuela

Masdevallia pileata Luer & Würstle

Distribution: Colombia

Masdevallia pinocchio Luer & Andreetta

Distribution: Ecuador

Masdevallia planadensis Luer & R.Escobar

Distribution: Colombia, Ecuador

Masdevallia plantaginea (Poepp'& Endl.) Cogn.
  Humboldtia plantaginea (Poepp & Endl.) Kuntze
  Masdevallia capillaris Luer
  Pleurothallis plantaginea (Poepp & Endl.) Lindl.
  Specklinia plantaginea Poepp & Endl.

Distribution: Ecuador, Peru

Masdevallia platyglossa Rchb.f.
  Lothiania bilabiata Kraenzl.
  Masdevallia bilabiata (Kraenzl.) Garay
  Rodrigoa bilabiata (Kraenzl.) Braas

Distribution: Colombia, Ecuador

Masdevallia pleurothalloides Luer

Distribution: Panama

Masdevallia plynophora Luer

Distribution: Peru

Masdevallia × polita Luer & Sijm

Distribution: Unknown

Masdevallia pollux Luer & Cloes

Distribution: Ecuador, Peru

Masdevallia polychroma Luer

Distribution: Ecuador

Masdevallia polysticta Rchb.f.
Masdevallia huebschiana Kraenzl.
Masdevallia spathulifolia Kraenzl.

Distribution: Ecuador, Peru

Masdevallia popowiana Königer & J.G.Weinm.bis

Distribution: Peru

Masdevallia porphyrea Luer

Distribution: Ecuador

Masdevallia portillae Luer & Andreetta

Distribution: Ecuador

Masdevallia posadae Luer & R.Escobar
Masdevallia sijmiana Königer

Distribution: Colombia, Peru

Masdevallia pozoi Königer

Distribution: Ecuador, Peru

Part II: Masdevallia

Masdevallia princeps Luer

Distribution: Peru

Masdevallia proboscoidea Luer & V.N.M.Rao

Distribution: Ecuador

Masdevallia prodigiosa Königer

Distribution: Peru

Masdevallia prolixa Luer

Distribution: Peru

Masdevallia prosartema Königer

Distribution: Peru

Masdevallia pteroglossa Schltr.
  Masdevallia xerophila F.Lehm & Kraenzl.

Distribution: Colombia

Masdevallia pulcherrima Luer & Andreetta

Distribution: Ecuador

Masdevallia pumila Poepp & Endl.
  Masdevallia filamentosa Kraenzl.
  Masdevallia grandiflora C.Schweinf.

Distribution: Bolivia, Colombia, Ecuador, Peru

Masdevallia purpurella Luer & R.Escobar

Distribution: Colombia

Masdevallia pyknosepala Luer & Cloes

Distribution: Peru

Masdevallia pyxis Luer

Distribution: Peru

Masdevallia quasimodo Luer

Distribution: Bolivia

Masdevallia racemosa Lindl.

Distribution: Colombia

Masdevallia rafaeliana Luer

Distribution: Costa Rica, Panama

Masdevallia receptrix Luer & R.Vasquez

Distribution: Bolivia

Masdevallia rechingeriana Kraenzl.

Distribution: Venezuela

Masdevallia recurvata Luer & Dalstrom

Distribution: Peru

Masdevallia regina Luer

Distribution: Peru

Masdevallia reichenbachiana Endres ex Rchb.f.
Masdevallia funebris Endres ex Kraenzl.
Masdevallia normannii hort.

Distribution: Costa Rica

Masdevallia renzii Luer

Distribution: Colombia

Masdevallia repanda Luer & Hirtz

Distribution: Ecuador

Masdevallia replicata Königer

Distribution: Peru

Part II: Masdevallia

Masdevallia revoluta Königer & J.Portilla

Distribution: Ecuador

Masdevallia rex Luer & Hirtz

Distribution: Ecuador

Masdevallia rhinophora Luer & R.Escobar

Distribution: Colombia

Masdevallia rhodehameliana Luer

Distribution: Peru

Masdevallia richardsoniana Luer

Distribution: Peru

Masdevallia ricii Luer & R.Vasquez

Distribution: Bolivia

Masdevallia rigens Luer
  Masdevallia stercorea Königer

Distribution: Peru

Masdevallia rimarima-alba Luer

Distribution: Peru

Masdevallia robusta Luer

Distribution: Ecuador

Masdevallia rodolfoi (Braas) Luer

Distribution: Peru

Masdevallia rolandorum Luer & Sijm

Distribution: Peru

Masdevallia rolfeana Kraenzl.

Distribution: Costa Rica

Masdevallia rosea Lindl.
Masdevallia echinata Luer & Andreetta

Distribution: Colombia, Ecuador

Masdevallia roseola Luer

Distribution: Ecuador

Masdevallia rubeola Luer & R.Vasquez

Distribution: Bolivia, Peru

Masdevallia rubiginosa Königer

Distribution: Ecuador, Peru

Masdevallia rufescens Königer
Masdevallia neglecta Königer

Distribution: Ecuador, Peru

Masdevallia rugulosa Königer

Distribution: Peru

Masdevallia saltatrix Rchb.f.

Distribution: Colombia

Masdevallia sanchezii Luer & Andreetta

Distribution: Ecuador

Masdevallia sanctae-fidei Kraenzl.
Masdevallia dispar Luer

Distribution: Colombia, Venezuela

Masdevallia sanctae-inesiae Luer & Malo

Distribution: Ecuador

135

Part II: Masdevallia

Masdevallia sanctae-rosae Kraenzl.
Pleurothallis sanctae-rosae (Kraenzl.) Garay

Distribution: Colombia

Masdevallia sanguinea Luer & Andreetta

Distribution: Ecuador

Masdevallia scabrilinguis Luer

Distribution: Costa Rica, Panama

Masdevallia scalpellifera Luer

Distribution: Ecuador

Masdevallia scandens Rolfe
Masdevallia buchtienii Schltr.

Distribution: Bolivia

Masdevallia sceptrum Rchb.f.
Masdevallia urostachya Rchb.f.

Distribution: Colombia, Venezuela

Masdevallia schizantha Kraenzl.

Distribution: Colombia

Masdevallia schizopetala Kraenzl.
Masdevallia morenoi Luer

Distribution: Bolivia, Colombia, Costa Rica, Panama

Masdevallia schizostigma Luer

Distribution: Peru

Masdevallia schlimii Linden ex Lindl.
Masdevallia polyantha Lindl.

Distribution: Colombia, Venezuela

Masdevallia schmidt-mummii Luer & R.Escobar

Distribution: Colombia

Masdevallia schoonenii Luer

Distribution: Peru

Masdevallia schroederiana Sander ex Veitch
   Masdevallia schroederae Boos (sphalm.)

Distribution: Costa Rica, Peru

Masdevallia schudelii Luer

Distribution: Ecuador

Masdevallia scitula Königer

Distribution: Peru

Masdevallia scobina Luer & R.Escobar

Distribution: Colombia

Masdevallia scopaea Luer & R.Vasquez

Distribution: Bolivia

Masdevallia segrex Luer & Hirtz

Distribution: Ecuador

Masdevallia segurae Luer & R.Escobar
   Rodrigoa segurae (Luer & R.Escobar) Braas

Distribution: Colombia

Masdevallia selenites Königer

Distribution: Peru

Masdevallia semiteres Luer & R.Escobar

Distribution: Peru

137

Part II: Masdevallia

Masdevallia × senghasiana Luer
Masdevallia × lueri Senghas

Distribution: Colombia

Masdevallia serendipita Luer & Teague

Distribution: Bolivia

Masdevallia sernae Luer & R.Escobar
Masdevallia carolloi Luer & Andreetta

Distribution: Colombia, Ecuador

Masdevallia sertula Luer & Andreetta

Distribution: Ecuador

Masdevallia setacea Luer & Malo

Distribution: Ecuador, Peru

Masdevallia setipes Schltr.

Distribution: Bolivia

Masdevallia siphonantha Luer

Distribution: Colombia

Masdevallia smallmaniana Luer

Distribution: Ecuador

Masdevallia soennemarkii Luer & Dalstrom

Distribution: Bolivia

Masdevallia solomonii Luer & R.Vasquez

Distribution: Bolivia

Masdevallia spilantha Königer

Distribution: Peru

Masdevallia × splendida Rchb.f.
  Masdevallia × parlatoreana Rchb.f.

Distribution: Peru

Masdevallia sprucei Rchb.f.

Distribution: Brazil, Venezuela

Masdevallia staaliana Luer & Hirtz

Distribution: Ecuador

Masdevallia stenorhynchos Kraenzl.

Distribution: Colombia

Masdevallia stigii Luer & Jost

Distribution: Ecuador

Masdevallia stirpis Luer

Distribution: Venezuela

Masdevallia strattoniana Luer & Hirtz

Distribution: Ecuador

Masdevallia striatella Rchb.f.
  Masdevallia chloracra Rchb.f.
  Masdevallia superflua Kraenzl.

Distribution: Costa Rica, Panama, Venezuela

Masdevallia strigosa Königer

Distribution: Ecuador

Masdevallia strobelii H.R.Sweet & Garay

Distribution: Ecuador

Masdevallia × strumella Luer

Distribution: Colombia

Part II: Masdevallia

Masdevallia strumifera Rchb.f.
Masdevallia chrysochaete F.Lehm.
Masdevallia maxillariiformis F.Lehm & Kraenzl.

Distribution: Colombia, Ecuador, Venezuela

Masdevallia strumosa P.Ortíz & C.E.Calderón

Distribution: Colombia

Masdevallia stumpflei Braas

Distribution: Peru

Masdevallia suinii Luer & Hirtz

Distribution: Ecuador

Masdevallia sulphurella Königer

Distribution: Peru

Masdevallia sumapazensis P.Ortíz

Distribution: Colombia

Masdevallia × synthesis Luer

Distribution: Venezuela

Masdevallia teaguei Luer
Jostia teaguei (Luer) Luer
Masdevallia braasii Mohr

Distribution: Ecuador

Masdevallia tentaculata Luer

Distribution: Ecuador

Masdevallia terborchii Luer

Distribution: Peru

Masdevallia theleura Luer

Distribution: Ecuador

Masdevallia thienii Dodson

Distribution: Colombia, Costa Rica, Ecuador, Panama

Masdevallia tinekeae Luer & R.Vasquez

Distribution: Bolivia

Masdevallia titan Luer

Distribution: Peru

Masdevallia tokachiorum Luer

Distribution: Panama

Masdevallia tonduzii Woolward

Distribution: Costa Rica, Panama

Masdevallia torta Rchb.f.

Distribution: Colombia

Masdevallia tovarensis Rchb.f.
  Masdevallia candida Klotzsch & H.Karst. ex Rchb.f.
  Masdevallia candida Linden non Klotzsch & H.Karst.ex Rchb.f.

Distribution: Venezuela

Masdevallia trautmanniana Luer & J.Portilla

Distribution: Ecuador

Masdevallia triangularis Lindl.

Distribution: Colombia, Ecuador, Venezuela

Masdevallia tricallosa Königer

Distribution: Peru

141

Part II: Masdevallia

Masdevallia tricycla Luer

Distribution: Ecuador

Masdevallia tridens Rchb.f.
Masdevallia jubar Luer & Malo

Distribution: Colombia, Ecuador

Masdevallia trifurcata Luer

Distribution: Ecuador

Masdevallia trigonopetala Kraenzl.

Distribution: Colombia, Ecuador

Masdevallia trochilus Linden & André
Masdevallia colibri Burb.
Masdevallia ephippium Rchb.f.

Distribution: Colombia, Ecuador, Peru

Masdevallia truncata Luer

Distribution: Ecuador

Masdevallia tsubotae Luer

Distribution: Colombia

Masdevallia tubata Schltr.

Distribution: Bolivia

Masdevallia tubuliflora Ames
Masdevallia ecaudata Schltr.

Distribution: Belize, Costa Rica, Guatemala, Nicaragua

Masdevallia tubulosa Lindl.
Masdevallia casta Kraenzl.
Masdevallia stenantha F.Lehm & Kraenzl.
Masdevallia syringodes Luer & Andreetta

Distribution: Colombia, Ecuador, Peru, Venezuela

Masdevallia uncifera Rchb.f.
Masdevallia chrysoneura F.Lehm & Kraenzl.
Masdevallia flaccida Kraenzl.
Masdevallia pastensis Kraenzl.

Distribution: Colombia, Ecuador

Masdevallia uniflora Ruiz & Pav.

Distribution: Peru

Masdevallia urceolaris Kraenzl.
Masdevallia kalbreyeri Rchb.f.ex Kraenzl.

Distribution: Colombia

Masdevallia ustulata Luer

Distribution: Colombia, Ecuador, Peru

Masdevallia utriculata Luer

Distribution: Panama

Masdevallia valenciae Luer & R.Escobar

Distribution: Colombia

Masdevallia vargasii C.Schweinf.
Masdevallia megaloglossa Luer & R.Escobar
Masdevallia richteri Pabst

Distribution: Bolivia, Brazil, Colombia, Ecuador, Peru

Masdevallia vasquezii Luer

Distribution: Bolivia

Masdevallia veitchiana Rchb.f.

Distribution: Peru

Masdevallia velella Luer

Distribution: Colombia

Part II: Masdevallia

Masdevallia velifera Rchb.f.
Masdevallia valifera Sphalm.
Masdevallia vilifera Sphalm.

Distribution: Colombia

Masdevallia venatoria Luer & Malo

Distribution: Ecuador

Masdevallia venezuelana H.R.Sweet

Distribution: Venezuela

Masdevallia ventricosa Schltr.

Distribution: Ecuador

Masdevallia ventricularia Rchb.f.

Distribution: Colombia, Ecuador

Masdevallia venus Luer & Hirtz

Distribution: Ecuador

Masdevallia venusta Schltr.

Distribution: Peru

Masdevallia verecunda Luer

Distribution: Venezuela

Masdevallia vexillifera Luer

Distribution: Peru

Masdevallia vidua Luer & Andreetta

Distribution: Ecuador

Masdevallia vieirana Luer & R.Escobar

Distribution: Colombia

Masdevallia villegasii Königer

Distribution: Colombia

Masdevallia virens Luer & Andreetta

Distribution: Ecuador

Masdevallia virgo-cuencae Luer & Andreetta

Distribution: Ecuador

Masdevallia vittatula Luer & R.Escobar

Distribution: Colombia, Ecuador

Masdevallia vomeris Luer

Distribution: Peru

Masdevallia wageneriana Linden ex Lindl.

Distribution: Venezuela

Masdevallia walteri Luer

Distribution: Costa Rica

Masdevallia weberbaueri Schltr.
   Masdevallia moyobambae Königer

Distribution: Ecuador, Peru

Masdevallla welischii Luer

Distribution: Peru

Masdevallia wendlandiana Rchb.f.
   Masdevallia rodrigueziana Mansf.
   Masdevallia ulei Schltr.
   Masdevallia yauaperyensis Barb.Rodr.

Distribution: Bolivia, Brazil, Colombia, Ecuador, Peru, Venezuela

Part II: Masdevallia

Masdevallia whiteana Luer

Distribution: Ecuador, Peru

Masdevallia × wubbenii Luer

Distribution: Venezuela

Masdevallia wuelfinghoffiana Luer & J.Portilla

Distribution: Ecuador

Masdevallia wuerstlei Luer

Distribution: Colombia

Masdevallia wurdackii C.Schweinf.

Distribution: Peru

Masdevallia xanthina Rchb.f.
 Masdevallia pallida (Woolward) Luer

Distribution: Colombia, Ecuador

Masdevallia xanthodactyla Rchb.f.

Distribution: Ecuador, Peru

Masdevallia ximenae Luer & Hirtz

Distribution: Ecuador

Masdevallia xylina Rchb.f.

Distribution: Colombia

Masdevallia yungasensis T.Hashim.
 Masdevallia calocodon Luer & R.Vasquez

Distribution: Bolivia

Masdevallia zahlbruckneri Kraenzl.
Masdevallia humilis Luer
Masdevallia olmosii Königer

Distribution: Bolivia, Colombia, Costa Rica, Ecuador, Panama

Masdevallia zamorensis Luer & J.Portilla

Distribution: Ecuador

Masdevallia zapatae Luer & R.Escobar

Distribution: Colombia

Masdevallia zebracea Luer

Distribution: Peru

Masdevallia zongoënsis Luer & Hirtz

Distribution: Bolivia

Masdevallia zumbae Luer

Distribution: Ecuador

Masdevallia zumbuehlerae Luer

Distribution: Ecuador

Masdevallia zygia Luer & Malo

Distribution: Ecuador

PART III: COUNTRY CHECKLIST
For the genera:

Aerides, *Coelogyne*, *Comparettia* and *Masdevallia*

TROISIÈME PARTIE: LISTE PAR PAYS
Pour les genre:

Aerides, *Coelogyne*, *Comparettia* et *Masdevallia*

PARTE III: LISTA POR PAÍSES
Para el genero:

Aerides, *Coelogyne*, *Comparettia* y *Masdevallia*

PART III: COUNTRY CHECKLIST FOR THE GENERA:
Aerides, Coelogyne, Comparettia and *Masdevallia*

TROISIÈME PARTIE: LISTE PAR PAYS POUR LES GENRE:
Aerides, Coelogyne, Comparettia et *Masdevallia*

PARTE III: LISTA POR PAISES PARA EL GENERO:
Aerides, Coelogyne, Comparettia y *Masdevallia*

BANGLADESH / BANGLADESH (LE) / BANGLADESH

Aerides multiflora Roxb.
Coelogyne nitida (Wall. ex D.Don) Lindl.
Coelogyne viscosa Rchb.f.

BELIZE / BELIZE (LE) / BELICE

Comparettia falcata Poepp. & Endl.
Masdevallia adamsii Luer
Masdevallia floribunda Lindl.
Masdevallia tubuliflora Ames

BHUTAN / BHOUTAN (LE) / BHUTÁN

Aerides multiflora Roxb.
Aerides odorata Lour.
Aerides rosea Lodd. ex Lindl. & Paxton
Coelogyne assamica Linden & Rchb.f.
Coelogyne barbata Griff.
Coelogyne corymbosa Lindl.
Coelogyne cristata Lindl.
Coelogyne fimbriata Lindl.
Coelogyne fuscescens
Coelogyne holochila P.F.Hunt & Summerh.
Coelogyne longipes Lindl.
Coelogyne nitida (Wall. ex D.Don) Lindl.
Coelogyne occultata
Coelogyne ovalis Lindl.
Coelogyne prolifera Lindl.
Coelogyne punctulata Lindl.
Coelogyne raizadae S.K.Jain & S.Das
Coelogyne schultesii S.K.Jain & S.Das
Coelogyne stricta (D.Don) Schltr.

BOLIVIA / BOLIVIE (LA) / BOLIVIA

Comparettia coccinea T.Hashim.
Masdevallia bangii Schltr.
Masdevallia bicolor Poepp. & Endl.
Masdevallia boliviensis Schltr.

150

Masdevallia brachyantha Schltr.
Masdevallia burianii Luer & Dalstrom
Masdevallia chaparensis T.Hashim.
Masdevallia chuspipatae Luer & Teague
Masdevallia cocapatae Luer, Teague & R.Vasquez
Masdevallia datura Luer & R.Vasquez
Masdevallia elachys Luer
Masdevallia exquisita Luer & Hirtz
Masdevallia frilehmannii Luer & R.Vasquez
Masdevallia gutierrezii Luer
Masdevallia heideri Königer
Masdevallia helenae Luer
Masdevallia infracta Lindl.
Masdevallia ishikoi Luer
Masdevallia isos Luer
Masdevallia lewisii Luer & R.Vasquez
Masdevallia luziae-mariae Luer & R.Vasquez
Masdevallia martineae Luer
Masdevallia mascarata Luer & R.Vasquez
Masdevallia menatoi Luer & R.Vasquez
Masdevallia minuta Lindl.
Masdevallia nebulina Luer
Masdevallia nitens Luer
Masdevallia notosibirica F.Maek. & T.Hashim.
Masdevallia omorenoi Luer & R.Vasquez
Masdevallia oreas Luer & R.Vasquez
Masdevallia paivaëana Rchb.f.
Masdevallia parvula Schltr.
Masdevallia phacopsis Luer & Dalstrom
Masdevallia picturata Rchb.f.
Masdevallia pumila Poepp. & Endl.
Masdevallia quasimodo Luer
Masdevallia receptrix Luer & R.Vasquez
Masdevallia ricii Luer & R.Vasquez
Masdevallia rubeola Luer & R.Vasquez
Masdevallia scandens Rolfe
Masdevallia schizopetala Kraenzl.
Masdevallia scopaea Luer & R.Vasquez
Masdevallia serendipita Luer & Teague
Masdevallia setipes Schltr.
Masdevallia soennemarkii Luer & Dalstrom
Masdevallia solomonii Luer & R.Vasquez
Masdevallia tinekeae Luer & R.Vasquez
Masdevallia tubata Schltr.
Masdevallia vargasii C.Schweinf.
Masdevallia vasquezii Luer
Masdevallia wendlandiana Rchb.f.
Masdevallia yungasensis T.Hashim.
Masdevallia zahlbruckneri Kraenzl.
Masdevallia zongoënsis Luer & Hirtz

BRAZIL / BRÉSIL (LE) / BRASIL (EL)

Comparettia coccinea Lindl.
Comparettia falcata Poepp. & Endl.
Masdevallia cuprea Lindl.
Masdevallia curtipes Barb.Rodr.
Masdevallia discoidea Luer & Würstle
Masdevallia infracta Lindl.
Masdevallia minuta Lindl.
Masdevallia norae Luer
Masdevallia obscurans (Luer) Luer
Masdevallia oscitans (Luer) Luer
Masdevallia sprucei Rchb.f.
Masdevallia vargasii C.Schweinf.
Masdevallia wendlandiana Rchb.f.

BRUNEI DARUSSALAM / BRUNÉI DARUSSALAM (LE) / BRUNEI DARUSSALAM

Coelogyne asperata Lindl.
Coelogyne bruneiensis de Vogel
Coelogyne craticulaelabris Carr
Coelogyne echinolabium de Vogel
Coelogyne exalata Ridl.
Coelogyne genuflexa Ames & C.Schweinf.
Coelogyne gibbifera J.J.Sm.
Coelogyne hirtella J.J.Sm.
Coelogyne incrassata (Blume) Lindl.
Coelogyne odoardii Schltr.
Coelogyne pandurata Lindl.
Coelogyne pulverula Teijsm. & Binn.
Coelogyne radioferens Ames & C.Schweinf.
Coelogyne rochussenii de Vriese
Coelogyne sanderiana Rchb.f.
Coelogyne septemcostata J.J.Sm.
Coelogyne swaniana Rolfe
Coelogyne venusta Rolfe
Coelogyne verrucosa S.E.C.Sierra

CAMBODIA / CAMBODGE (LE) / CAMBOYA

Aerides falcata Lindl. & Paxton
Aerides houlletiana Rchb.f.
Aerides multiflora Roxb.
Aerides odorata Lour.
Coelogyne fimbriata Lindl.
Coelogyne trinervis Lindl.

CHINA / CHINE (LA) / CHINA

Aerides flabellata Rolfe ex Downie
Aerides odorata Lour.
Aerides rosea Lodd. ex Lindl. & Paxton
Coelogyne assamica Linden & Rchb.f.
Coelogyne barbata Griff.
Coelogyne calcicola Kerr
Coelogyne corymbosa Lindl.
Coelogyne cristata Lindl.
Coelogyne esquirolii Schltr.
Coelogyne fimbriata Lindl.
Coelogyne flaccida Lindl.
Coelogyne fuscescens
Coelogyne gongshanensis H.Li ex S.C.Chen
Coelogyne griffithii Hook.f.
Coelogyne holochila P.F.Hunt & Summerh.
Coelogyne leucantha W.W.Sm.
Coelogyne lockii Aver.
Coelogyne longipes Lindl.
Coelogyne malipoensis Z.H.Tsi
Coelogyne nitida (Wall. ex D.Don) Lindl.
Coelogyne occultata Hook.f.
Coelogyne ovalis Lindl.
Coelogyne prolifera Lindl.
Coelogyne pulchella Rolfe
Coelogyne punctulata Lindl.
Coelogyne raizadae S.K.Jain & S.Das
Coelogyne rigida C.S.P.Parish & Rchb.f.
Coelogyne sanderae Kraenzl. ex O'Brien
Coelogyne schultesii S.K.Jain & S.Das
Coelogyne stricta (D.Don) Schltr.
Coelogyne suaveolens (Lindl.) Hook.f.
Coelogyne taronensis Hand.-Mazz.
Coelogyne viscosa Rchb.f.
Coelogyne zhenkangensis S.C.Chen & K.Y.Lang

COLOMBIA / COLOMBIE (LA) / COLOMBIA

Comparettia coccinea Lindl.
Comparettia falcata Poepp. & Endl.
Comparettia ignea P.Ortíz
Comparettia macroplectron Rchb.f. & Triana
Masdevallia aenigma Luer & R.Escobar
Masdevallia aguirrei Luer & R.Escobar
Masdevallia akemiana Königer & Sijm
Masdevallia alismifolia Kraenzl.
Masdevallia × alvaroi Luer & R.Escobar
Masdevallia amanda Rchb.f. & Warsz.
Masdevallia angulata Rchb.f.
Masdevallia angulifera Rchb.f. ex Kraenzl.
Masdevallia anisomorpha Garay
Masdevallia × anthina Rchb.f.

Masdevallia apparitio Luer & R.Escobar
Masdevallia arangoi Luer & R.Escobar
Masdevallia arminii Linden & Rchb.f.
Masdevallia assurgens Luer & R.Escobar
Masdevallia bicolor Poepp. & Endl.
Masdevallia bonplandii Rchb.f.
Masdevallia brockmuelleri Luer
Masdevallia buccinator Rchb.f. & Warsz.
Masdevallia cacodes Luer & R.Escobar
Masdevallia caesia Roezl
Masdevallia campyloglossa Rchb.f.
Masdevallia caudata Lindl.
Masdevallia caudivolvula Kraenzl.
Masdevallia cerastes Luer & R.Escobar
Masdevallia chimboënsis Kraenzl.
Masdevallia clandestina Luer & R.Escobar
Masdevallia coccinea Linden ex Lindl.
Masdevallia coriacea Lindl.
Masdevallia corniculata Rchb.f.
Masdevallia crescenticola F.Lehm. & Kraenzl.
Masdevallia cucullata Lindl.
Masdevallia densiflora Schltr.
Masdevallia discolor Luer & R.Escobar
Masdevallia dryada Luer & R.Escobar
Masdevallia elephanticeps Rchb.f. & Warsz.
Masdevallia encephala Luer & R.Escobar
Masdevallia estradae Rchb.f.
Masdevallia expansa Rchb.f.
Masdevallia falcago Rchb.f.
Masdevallia fasciata Rchb.f.
Masdevallia filaria Luer & R.Escobar
Masdevallia floribunda Lindl.
Masdevallia foetens Luer & R.Escobar
Masdevallia fragrans Woolward
Masdevallia gargantua Rchb.f.
Masdevallia geminiflora P.Ortíz
Masdevallia gilbertoi Luer & R.Escobar
Masdevallia hercules Luer & Andreetta
Masdevallia herradurae F.Lehm. & Kraenzl.
Masdevallia heteroptera Rchb.f.
Masdevallia hians Linden & Rchb.f.
Masdevallia hieroglyphica Rchb.f.
Masdevallia hortensis Luer & R.Escobar
Masdevallia hubeinii Luer & Würstle
Masdevallia hylodes Luer & R.Escobar
Masdevallia ignea Rchb.f.
Masdevallia impostor Luer & R.Escobar
Masdevallia indecora Luer & R.Escobar
Masdevallia klabochiorum Rchb.f.
Masdevallia laevis Lindl.
Masdevallia leontoglossa Rchb.f.
Masdevallia leptoura Luer
Masdevallia × ligiae Luer & R.Escobar
Masdevallia livingstoneana Roezl

Masdevallia ludibunda Rchb.f.
Masdevallia ludibundella Luer & R.Escobar
Masdevallia macrogenia (Arango) Luer & R.Escobar
Masdevallia macroglossa Rchb.f.
Masdevallia macrura Rchb.f.
Masdevallia mandarina (Luer & R.Escobar) Luer
Masdevallia marthae Luer & R.Escobar
Masdevallia mastodon Rchb.f.
Masdevallia medusa Luer & R.Escobar
Masdevallia mejiana Garay
Masdevallia melanoxantha Linden & Rchb.f.
Masdevallia meleagris Lindl.
Masdevallia minuta Lindl.
Masdevallia misasii Braas
Masdevallia molossus Rchb.f.
Masdevallia mooreana Rchb.f.
Masdevallia mutica Luer & R.Escobar
Masdevallia × mystica Luer
Masdevallia nidifica Rchb.f.
Masdevallia niesseniae Luer
Masdevallia nivea (Luer & R.Escobar) Luer & R.Escobar
Masdevallia norae Luer
Masdevallia odontocera Luer & R.Escobar
Masdevallia os-draconis Luer & R.Escobar
Masdevallia oscarii Luer & R.Escobar
Masdevallia pachyantha Rchb.f.
Masdevallia pachysepala (Rchb.f.) Luer
Masdevallia pardina Rchb.f.
Masdevallia parvula Schltr.
Masdevallia pastinata Luer
Masdevallia peristeria Rchb.f.
Masdevallia pescadoënsis Luer & R.Escobar
Masdevallia picturata Rchb.f.
Masdevallia pileata Luer & Würstle
Masdevallia planadensis Luer & R.Escobar
Masdevallia platyglossa Rchb.f.
Masdevallia posadae Luer & R.Escobar
Masdevallia pteroglossa Schltr.
Masdevallia pumila Poepp. & Endl.
Masdevallia purpurella Luer & R.Escobar
Masdevallia racemosa Lindl.
Masdevallia renzii Luer
Masdevallia rhinophora Luer & R.Escobar
Masdevallia rosea Lindl.
Masdevallia saltatrix Rchb.f.
Masdevallia sanctae-fidei Kraenzl.
Masdevallia sanctae-rosae Kraenzl.
Masdevallia sceptrum Rchb.f.
Masdevallia schizantha Kraenzl.
Masdevallia schizopetala Kraenzl.
Masdevallia schlimii Linden ex Lindl.
Masdevallia schmidt-mummii Luer & R.Escobar
Masdevallia scobina Luer & R.Escobar
Masdevallia segurae Luer & R.Escobar

Masdevallia × senghasiana Luer
Masdevallia sernae Luer & R.Escobar
Masdevallia siphonantha Luer
Masdevallia stenorhynchos Kraenzl.
Masdevallia × strumella Luer
Masdevallia strumifera Rchb.f.
Masdevallia strumosa P.Ortíz & C.E.Calderón
Masdevallia sumapazensis P.Ortíz
Masdevallia thienii Dodson
Masdevallia torta Rchb.f.
Masdevallia triangularis Lindl.
Masdevallia tridens Rchb.f.
Masdevallia trigonopetala Kraenzl.
Masdevallia trochilus Linden & André
Masdevallia tsubotae Luer
Masdevallia tubulosa Lindl.
Masdevallia uncifera Rchb.f.
Masdevallia urceolaris Kraenzl.
Masdevallia ustulata Luer
Masdevallia valenciae Luer & R.Escobar
Masdevallia vargasii C.Schweinf.
Masdevallia velella Luer
Masdevallia velifera Rchb.f.
Masdevallia ventricularia Rchb.f.
Masdevallia vieirana Luer & R.Escobar
Masdevallia villegasii Königer
Masdevallia vittatula Luer & R.Escobar
Masdevallia wendlandiana Rchb.f.
Masdevallia wuerstlei Luer
Masdevallia xanthina Rchb.f.
Masdevallia xylina Rchb.f.
Masdevallia zahlbruckneri Kraenzl.
Masdevallia zapatae Luer & R.Escobar

COSTA RICA / COSTA RICA (LE) / COSTA RICA

Comparettia falcata Poepp. & Endl.
Comparettia falcata Poepp. & Endl.
Masdevallia attenuata Rchb.f.
Masdevallia calura Rchb.f.
Masdevallia chasei Luer
Masdevallia chontalensis Rchb.f.
Masdevallia cupularis Rchb.f.
Masdevallia demissa Rchb.f.
Masdevallia epallax Königer
Masdevallia flaveola Rchb.f.
Masdevallia floribunda Lindl.
Masdevallia fulvescens Rolfe
Masdevallia lankesterana Luer
Masdevallia lata Rchb.f.
Masdevallia laucheana Kraenzl.
Masdevallia livingstoneana Roezl
Masdevallia marginella Rchb.f.

Masdevallia molossoides Kraenzl.
Masdevallia nidifica Rchb.f.
Masdevallia picturata Rchb.f.
Masdevallia rafaeliana Luer
Masdevallia reichenbachiana Endres ex Rchb.f.
Masdevallia rolfeana Kraenzl.
Masdevallia scabrilinguis Luer
Masdevallia schizopetala Kraenzl.
Masdevallia schroederiana Sander ex Veitch
Masdevallia striatella Rchb.f.
Masdevallia thienii Dodson
Masdevallia tonduzii Woolward
Masdevallia tubuliflora Ames
Masdevallia walteri Luer
Masdevallia zahlbruckneri Kraenzl.

CUBA / CUBA / CUBA

Comparettia falcata Poepp. & Endl.

ECUADOR / EQUATEUR (L') / ECUADOR (EL)

Comparettia speciosa Rchb.f.
Comparettia × maloi I.Bock
Masdevallia abbreviata Rchb.f.
Masdevallia acaroi Luer & Hirtz
Masdevallia acrochordonia Rchb.f.
Masdevallia adrianae Luer
Masdevallia agaster Luer
Masdevallia albella Luer & Teague
Masdevallia alexandri Luer
Masdevallia amaluzae Luer & Malo
Masdevallia amanda Rchb.f. & Warsz.
Masdevallia ametroglossa Luer & Hirtz
Masdevallia amoena Luer
Masdevallia ampullacea Luer & Andreetta
Masdevallia anceps Luer & Hirtz
Masdevallia andreettaeana Luer
Masdevallia anemone Luer
Masdevallia anfracta Königer & J.Portilla
Masdevallia angulata Rchb.f.
Masdevallia aphanes Königer
Masdevallia aptera Luer & L.O'Shaughn.
Masdevallia ariasii Luer
Masdevallia attenuata Rchb.f.
Masdevallia aurea Luer
Masdevallia bangii Schltr.
Masdevallia barrowii Luer
Masdevallia belua Königer & D'Aless.
Masdevallia berthae Luer & Andreetta
Masdevallia bicolor Poepp. & Endl.
Masdevallia bicornis Luer

Masdevallia bonplandii Rchb.f.
Masdevallia bottae Luer & Andreetta
Masdevallia bourdetteana Luer
Masdevallia brachyura F.Lehm. & Kraenzl.
Masdevallia brenneri Luer
Masdevallia bucculenta Luer & Hirtz
Masdevallia bulbophyllopsis Kraenzl.
Masdevallia calagrasalis Luer
Masdevallia calocalix Luer
Masdevallia campyloglossa Rchb.f.
Masdevallia carmenensis Luer & Malo
Masdevallia carruthersiana F.Lehm. & Kraenzl.
Masdevallia chaetostoma Luer
Masdevallia chimboënsis Kraenzl.
Masdevallia citrinella Luer & Malo
Masdevallia colossus Luer
Masdevallia condorensis Luer & Hirtz
Masdevallia constricta Poepp. & Endl.
Masdevallia corazonica Schltr.
Masdevallia corderoana F.Lehm. & Kraenzl.
Masdevallia coriacea Lindl.
Masdevallia corniculata Rchb.f.
Masdevallia crassicaudis Luer & J.Portilla
Masdevallia crescenticola F.Lehm. & Kraenzl.
Masdevallia cretata Luer
Masdevallia cucullata Lindl.
Masdevallia cuprea Lindl.
Masdevallia cylix Luer & Malo
Masdevallia dalessandroi Luer
Masdevallia dalstroemii Luer
Masdevallia decumana Königer
Masdevallia deformis Kraenzl.
Masdevallia delhierroi Luer & Hirtz
Masdevallia delphina Luer
Masdevallia deniseana Luer & J.Portilla
Masdevallia descendens Luer & Andreetta
Masdevallia dimorphotricha Luer & Hirtz
Masdevallia don-quijote Luer & Andreetta
Masdevallia dorisiae Luer
Masdevallia draconis Luer & Andreetta
Masdevallia dreisei Luer
Masdevallia dura Luer
Masdevallia dynastes Luer
Masdevallia ejiriana Luer & J.Portilla
Masdevallia empusa Luer
Masdevallia eucharis Luer
Masdevallia eurynogaster Luer & Andreetta
Masdevallia excelsior Luer & Andreetta
Masdevallia expers Luer & Andreetta
Masdevallia figueroae Luer
Masdevallia filaria Luer & R.Escobar
Masdevallia fractiflexa F.Lehm. & Kraenzl.
Masdevallia geminiflora P.Ortíz
Masdevallia glandulosa Königer

Masdevallia glomerosa Luer & Andreetta
Masdevallia gnoma Sweet
Masdevallia goliath Luer & Andreetta
Masdevallia graminea Luer
Masdevallia guerrieroi Luer & Andreetta
Masdevallia guttulata Rchb.f.
Masdevallia hartmanii Luer
Masdevallia helgae Königer & J.Portilla
Masdevallia henniae Luer & Dalstrom
Masdevallia hercules Luer & Andreetta
Masdevallia hirtzii Luer & Andreetta
Masdevallia hydrae Luer
Masdevallia hystrix Luer & Hirtz
Masdevallia impostor Luer & R.Escobar
Masdevallia ingridiana Luer & J.Portilla
Masdevallia instar Luer & Andreetta
Masdevallia josei Luer
Masdevallia klabochiorum Rchb.f.
Masdevallia laevis Lindl.
Masdevallia lamia Luer & Hirtz
Masdevallia lamprotyria Königer
Masdevallia lappifera Luer & Hirtz
Masdevallia leathersii Luer
Masdevallia lehmannii Rchb.f.
Masdevallia lenae Luer & Hirtz
Masdevallia leonardoi Luer
Masdevallia leptoura Luer
Masdevallia leucantha F.Lehm. & Kraenzl.
Masdevallia lilacina Königer
Masdevallia limax Luer
Masdevallia lintricula Königer
Masdevallia loui Luer & Dalstrom
Masdevallia lynniana Luer
Masdevallia macropus F.Lehm. & Kraenzl.
Masdevallia mallii Luer
Masdevallia maloi Luer
Masdevallia manchinazae Luer & Andreetta
Masdevallia mandarina (Luer & R.Escobar) Luer
Masdevallia manta Königer & Sijm
Masdevallia martiniana Luer
Masdevallia mataxa Königer & H.Mend.
Masdevallia maxilimax (Luer) Luer
Masdevallia mayaycu Luer & Andreetta
Masdevallia medinae Luer & J.Portilla
Masdevallia melanoglossa Luer
Masdevallia mendozae Luer
Masdevallia mentosa Luer
Masdevallia merinoi Luer & J.Portilla
Masdevallia microsiphon Luer
Masdevallia midas Luer
Masdevallia milagroi Luer & Hirtz
Masdevallia minuta Lindl.
Masdevallia × monicana Luer
Masdevallia morochoi Luer & Andreetta

Masdevallia murex Luer
Masdevallia naranjapatae Luer
Masdevallia newmaniana Luer & Teague
Masdevallia nidifica Rchb.f.
Masdevallia nigricans Königer & Sijm
Masdevallia nikoleana Luer & J.Portilla
Masdevallia norops Luer & Andreetta
Masdevallia odontopetala Luer
Masdevallia ophioglossa Rchb.f.
Masdevallia os-viperae Luer & Andreetta
Masdevallia ova-avis Luer
Masdevallia pachyura Rchb.f.
Masdevallia panguiënsis Luer & Andreetta
Masdevallia pantomima Luer & Hirtz
Masdevallia papillosa Luer
Masdevallia paquishae Luer & Hirtz
Masdevallia pardina Rchb.f.
Masdevallia parvula Schltr.
Masdevallia patchicutzae Luer & Hirtz
Masdevallia patriciana Luer
Masdevallia patula Luer & Malo
Masdevallia peristeria Rchb.f.
Masdevallia persicina Luer
Masdevallia picta Luer
Masdevallia picturata Rchb.f.
Masdevallia pinocchio Luer & Andreetta
Masdevallia planadensis Luer & R.Escobar
Masdevallia plantaginea (Poepp. & Endl.) Cogn.
Masdevallia platyglossa Rchb.f.
Masdevallia pollux Luer & Cloes
Masdevallia polychroma Luer
Masdevallia polysticta Rchb.f.
Masdevallia porphyrea Luer
Masdevallia portillae Luer & Andreetta
Masdevallia pozoi Königer
Masdevallia proboscoidea Luer & V.N.M.Rao
Masdevallia pulcherrima Luer & Andreetta
Masdevallia pumila Poepp. & Endl.
Masdevallia repanda Luer & Hirtz
Masdevallia revoluta Königer & J.Portilla
Masdevallia rex Luer & Hirtz
Masdevallia robusta Luer
Masdevallia rosea Lindl.
Masdevallia roseola Luer
Masdevallia rubiginosa Königer
Masdevallia rufescens Königer
Masdevallia sanchezii Luer & Andreetta
Masdevallia sanctae-inesiae Luer & Malo
Masdevallia sanguinea Luer & Andreetta
Masdevallia scalpellifera Luer
Masdevallia schudelii Luer
Masdevallia segrex Luer & Hirtz
Masdevallia sernae Luer & R.Escobar
Masdevallia sertula Luer & Andreetta

Masdevallia setacea Luer & Malo
Masdevallia smallmaniana Luer
Masdevallia staaliana Luer & Hirtz
Masdevallia stigii Luer & Jost
Masdevallia strattoniana Luer & Hirtz
Masdevallia strigosa Königer
Masdevallia strobelii H.R.Sweet & Garay
Masdevallia strumifera Rchb.f.
Masdevallia suinii Luer & Hirtz
Masdevallia teaguei Luer
Masdevallia tentaculata Luer
Masdevallia theleura Luer
Masdevallia thienii Dodson
Masdevallia trautmanniana Luer & J.Portilla
Masdevallia triangularis Lindl.
Masdevallia tricycla Luer
Masdevallia tridens Rchb.f.
Masdevallia trifurcata Luer
Masdevallia trigonopetala Kraenzl.
Masdevallia trochilus Linden & André
Masdevallia truncata Luer
Masdevallia tubulosa Lindl.
Masdevallia uncifera Rchb.f.
Masdevallia ustulata Luer
Masdevallia vargasii C.Schweinf.
Masdevallia venatoria Luer & Malo
Masdevallia ventricosa Schltr.
Masdevallia ventricularia Rchb.f.
Masdevallia venus Luer & Hirtz
Masdevallia vidua Luer & Andreetta
Masdevallia virens Luer & Andreetta
Masdevallia virgo-cuencae Luer & Andreetta
Masdevallia vittatula Luer & R.Escobar
Masdevallia weberbaueri Schltr.
Masdevallia wendlandiana Rchb.f.
Masdevallia whiteana Luer
Masdevallia wuelfinghoffiana Luer & J.Portilla
Masdevallia xanthina Rchb.f.
Masdevallia xanthodactyla Rchb.f.
Masdevallia ximenae Luer & Hirtz
Masdevallia zahlbruckneri Kraenzl.
Masdevallia zamorensis Luer & J.Portilla
Masdevallia zumbae Luer
Masdevallia zumbuehlerae Luer
Masdevallia zygia Luer & Malo

FIJI / FIDJI (LES) / FIJI

Coelogyne lycastoides F.Muell. & Kraenzl.
Coelogyne macdonaldii F.Muell. & Kraenzl.

**FRENCH GUIANA, OVERSEAS DEPARTMENT OF FRANCE /
GUYANE FRANÇAISE, TERRITOIRE D'OUTRE-MER DE LA FRANCE
/ GUYANA FRANCESA, TERRITORIO DE ULTRAMAR DE FRANCIA**

Masdevallia cuprea Lindl.
Masdevallia lansbergii Rchb.f.
Masdevallia minuta Lindl.

**GUAM, DEPENDANT TERRITORY OF THE UNITED STATES OF
AMERICA / GUAM, TERRITOIRE DÉPENDANT DES ETATS-UNIS
D'AMÉRIQUE / GUAM, TERRITORIO DEPENDIENTE DE LOS
ESTADOS UNIDOS DE AMÉRICA**

Coelogyne guamensis Ames

GUATEMALA / GUATEMALA (LE) / GUATEMALA

Comparettia falcata Poepp. & Endl.
Masdevallia chontalensis Rchb.f.
Masdevallia floribunda Lindl.
Masdevallia tubuliflora Ames

GUYANA / GUYANA (LE) / GUYANA

Masdevallia guayanensis Lindl. ex Benth.
Masdevallia minuta Lindl.
Masdevallia picturata Rchb.f.

HONDURAS / HONDURAS (LE) / HONDURAS

Comparettia falcata Poepp. & Endl.
Masdevallia floribunda Lindl.

INDIA / INDE (L') / INDIA (LA)

Aerides crassifolia C.S.P.Parish ex Burb.
Aerides crispa Lindl.
Aerides emericii Rchb.f.
Aerides falcata Lindl. & Paxton
Aerides maculosa Lindl.
Aerides mcmorlandii B.S.Williams
Aerides multiflora Roxb.
Aerides odorata Lour.
Aerides ringens (Lindl.) C.E.C.Fisch.
Aerides rosea Lodd. ex Lindl. & Paxton
Coelogyne albolutea Rolfe
Coelogyne assamica Linden & Rchb.f.
Coelogyne barbata Griff.
Coelogyne breviscapa Lindl.
Coelogyne corymbosa Lindl.

Coelogyne cristata Lindl.
Coelogyne fimbriata Lindl.
Coelogyne flaccida Lindl.
Coelogyne fuliginosa Lodd. ex Hook.
Coelogyne fuscescens
Coelogyne ghatakii T.K. Paul, S.K. Basu & M.C. Biswas
Coelogyne glandulosa
Coelogyne griffithii Hook.f.
Coelogyne hajrae Phukan
Coelogyne hitendrae S.Das & S.K.Jain
Coelogyne holochila P.F.Hunt & Summerh.
Coelogyne longeciliata Teijsm. & Binn.
Coelogyne longipes Lindl.
Coelogyne micrantha Lindl.
Coelogyne mossiae Rolfe
Coelogyne nervosa A.Rich.
Coelogyne nitida (Wall. ex D.Don) Lindl.
Coelogyne occultata Hook.f.
Coelogyne odoratissima Lindl.
Coelogyne ovalis Lindl.
Coelogyne pendula Summerh. ex Parry
Coelogyne prolifera Lindl.
Coelogyne punctulata Lindl.
Coelogyne quadratiloba Gagnep.
Coelogyne raizadae S.K.Jain & S.Das
Coelogyne rigida C.S.P.Parish & Rchb.f.
Coelogyne schultesii S.K.Jain & S.Das
Coelogyne stricta (D.Don) Schltr.
Coelogyne suaveolens (Lindl.) Hook.f.
Coelogyne tomentosa Lindl.
Coelogyne viscosa Rchb.f.
Coelogyne zeylanica Hook.f.

INDONESIA / INDONÉSIE (L') / INDONESIA

Aerides falcata Lindl. & Paxton
Aerides inflexa Teijsm. & Binn.
Aerides odorata Lour.
Aerides thibautiana Rchb.f.
Aerides timorana Miq.
Coelogyne asperata Lindl.
Coelogyne albobrunnea J.J.Sm.
Coelogyne beccarii Rchb.f.
Coelogyne bicamerata J.J.Sm.
Coelogyne borneensis Rolfe
Coelogyne brachygyne J.J.Sm.
Coelogyne buennemeyeri J.J.Sm.
Coelogyne calcarata J.J.Sm.
Coelogyne caloglossa Schltr.
Coelogyne carinata Rolfe
Coelogyne celebensis J.J.Sm.
Coelogyne chlorophaea Schltr.
Coelogyne chrysotropis Schltr.

Coelogyne clemensii Ames & C.Schweinf.
Coelogyne compressicaulis Ames & C.Schweinf.
Coelogyne concinna Ridl.
Coelogyne contractipetala J.J.Sm.
Coelogyne crassiloba J.J.Sm.
Coelogyne craticulaelabris Carr
Coelogyne cumingii Lindl.
Coelogyne cuprea H.Wendl. & Kraenzl.
Coelogyne distans J.J.Sm.
Coelogyne echinolabium de Vogel
Coelogyne endertii J.J.Sm.
Coelogyne flexuosa Rolfe
Coelogyne foerstermannii Rchb.f.
Coelogyne fonstenebrarum P.O'Byrne
Coelogyne formosa Schltr.
Coelogyne fragrans Schltr.
Coelogyne fuerstenbergiana Schltr.
Coelogyne fuliginosa Lodd. ex Hook.
Coelogyne genuflexa Ames & C.Schweinf.
Coelogyne gibbifera J.J.Sm.
Coelogyne harana J.J.Sm.
Coelogyne hirtella J.J.Sm.
Coelogyne imbricans J.J.Sm.
Coelogyne incrassata (Blume) Lindl.
Coelogyne integra Schltr.
Coelogyne kelamensis J.J.Sm.
Coelogyne kemiriensis J.J.Sm.
Coelogyne kinabaluensis Ames & C.Schweinf.
Coelogyne lacinulosa J.J.Sm.
Coelogyne latiloba de Vogel
Coelogyne longibulbosa Ames & C.Schweinf.
Coelogyne longifolia (Blume) Lindl.
Coelogyne longpasiaensis J.J.Wood & C.L.Chan
Coelogyne malintangensis J.J.Sm.
Coelogyne mayeriana Rchb.f.
Coelogyne miniata (Blume) Lindl.
Coelogyne monilirachis Carr
Coelogyne monticola J.J.Sm.
Coelogyne motleyi Rolfe ex J.J.Wood, D.A.Clayton & C.L.Chan
Coelogyne moultonii J.J.Sm.
Coelogyne multiflora Schltr.
Coelogyne naja J.J.Sm.
Coelogyne odoardii Schltr.
Coelogyne padangensis J.J.Sm. & Schltr.
Coelogyne pallens Ridl.
Coelogyne pandurata Lindl.
Coelogyne papillosa Ridl. ex Stapf
Coelogyne peltastes Rchb.f.
Coelogyne pholidotoides J.J.Sm.
Coelogyne planiscapa
Coelogyne plicatissima Ames & C.Schweinf.
Coelogyne prasina Ridl.
Coelogyne pulverula Teijsm. & Binn.
Coelogyne radicosa Ridl.

Coelogyne radioferens Ames & C.Schweinf.
Coelogyne rhabdobulbon Schltr.
Coelogyne rigidiformis Ames & C.Schweinf.
Coelogyne rochussenii de Vriese
Coelogyne rumphii Lindl.
Coelogyne rupicola Carr
Coelogyne salmonicolor Rchb.f.
Coelogyne sanderiana Rchb.f.
Coelogyne septemcostata J.J.Sm.
Coelogyne speciosa (Blume) Lindl.
Coelogyne squamulosa J.J.Sm.
Coelogyne steenisii J.J.Sm.
Coelogyne stenobulbon Schltr.
Coelogyne swaniana Rolfe
Coelogyne tenompokensis Carr
Coelogyne tenuis Rolfe
Coelogyne testacea Lindl.
Coelogyne tiomanensis M.R.Hend.
Coelogyne tomentosa Lindl.
Coelogyne trilobulata J.J.Sm.
Coelogyne trinervis Lindl.
Coelogyne triuncialis P.O'Byrne & J.J.Verm.
Coelogyne tumida J.J.Sm.
Coelogyne undatialata J.J.Sm.
Coelogyne veitchii Rolfe
Coelogyne vermicularis J.J.Sm.
Coelogyne xyrekes Ridl.
Coelogyne zurowetzii Carr

LAO PEOPLE'S DEMOCRATIC REPUBLIC (THE) / RÉPUBLIQUE DÉMOCRATIQUE POPULAIRE LAO (LA) / REPÚBLICA DEMOCRÁTICA POPULAR LAO (LA)

Aerides crassifolia C.S.P.Parish ex Burb.
Aerides falcata Lindl. & Paxton
Aerides flabellata Rolfe ex Downie
Aerides houlletiana Rchb.f.
Aerides multiflora Roxb.
Aerides odorata Lour.
Aerides rosea Lodd. ex Lindl. & Paxton
Coelogyne assamica Linden & Rchb.f.
Coelogyne calcicola Kerr
Coelogyne cumingii Lindl.
Coelogyne fimbriata Lindl.
Coelogyne flaccida Lindl.
Coelogyne fuscescens
Coelogyne longipes Lindl.
Coelogyne nitida (Wall. ex D.Don) Lindl.
Coelogyne pallens Ridl.
Coelogyne raizadae S.K.Jain & S.Das
Coelogyne schultesii S.K.Jain & S.Das
Coelogyne trinervis Lindl.
Coelogyne viscosa Rchb.f.

MALAYSIA / MALAISIE (LA) / MALASIA

Aerides falcata Lindl. & Paxton
Aerides krabiensis Seidenf.
Aerides multiflora Roxb.
Aerides odorata Lour.
Aerides sukauensis Shim
Coelogyne acutilabium de Vogel
Coelogyne anceps Hook.f.
Coelogyne albobrunnea J.J.Sm.
Coelogyne asperata Lindl.
Coelogyne chanii Gravend. & de Vogel
Coelogyne clemensii Ames & C.Schweinf.
Coelogyne compressicaulis Ames & C.Schweinf.
Coelogyne crassiloba J.J.Sm.
Coelogyne craticulaelabris Carr
Coelogyne cumingii Lindl.
Coelogyne cuprea H.Wendl. & Kraenzl.
Coelogyne distans J.J.Sm.
Coelogyne dulitensis Carr
Coelogyne echinolabium de Vogel
Coelogyne endertii J.J.Sm.
Coelogyne exalata Ridl.
Coelogyne fimbriata Lindl.
Coelogyne flexuosa Rolfe
Coelogyne foerstermannii Rchb.f.
Coelogyne genuflexa Ames & C.Schweinf.
Coelogyne gibbifera J.J.Sm.
Coelogyne harana J.J.Sm.
Coelogyne hirtella J.J.Sm.
Coelogyne imbricans J.J.Sm.
Coelogyne incrassata (Blume) Lindl.
Coelogyne judithiae P.Taylor
Coelogyne kaliana P.J.Cribb
Coelogyne kelamensis J.J.Sm.
Coelogyne kinabaluensis Ames & C.Schweinf.
Coelogyne latiloba de Vogel
Coelogyne longibulbosa Ames & C.Schweinf.
Coelogyne longpasiaensis J.J.Wood & C.L.Chan
Coelogyne marthae S.E.C.Sierra
Coelogyne mayeriana Rchb.f.
Coelogyne monilirachis Carr
Coelogyne motleyi Rolfe ex J.J.Wood,D.A.Clayton & C.L.Chan
Coelogyne moultonii J.J.Sm.
Coelogyne muluensis J.J.Wood
Coelogyne naja J.J.Sm.
Coelogyne obtusifolia Carr
Coelogyne odoardii Schltr.
Coelogyne pallens Ridl.
Coelogyne pandurata Lindl.
Coelogyne papillosa Ridl. ex Stapf
Coelogyne peltastes Rchb.f.
Coelogyne pholidotoides J.J.Sm.
Coelogyne planiscapa Carr

Coelogyne plicatissima Ames & C.Schweinf.
Coelogyne prasina Ridl.
Coelogyne pulverula Teijsm. & Binn.
Coelogyne radicosa Ridl.
Coelogyne radioferens Ames & C.Schweinf.
Coelogyne renae Gravend. & de Vogel
Coelogyne rhabdobulbon Schltr.
Coelogyne rigida C.S.P.Parish & Rchb.f.
Coelogyne rigidiformis Ames & C.Schweinf.
Coelogyne rochussenii de Vriese
Coelogyne rupicola Carr
Coelogyne sanderiana Rchb.f.
Coelogyne septemcostata J.J.Sm.
Coelogyne squamulosa J.J.Sm.
Coelogyne stenochila Hook.f.
Coelogyne swaniana Rolfe
Coelogyne tenompokensis Carr
Coelogyne testacea Lindl.
Coelogyne tiomanensis M.R.Hend.
Coelogyne tomentosa Lindl.
Coelogyne trinervis Lindl.
Coelogyne velutina de Vogel
Coelogyne venusta Rolfe
Coelogyne vermicularis J.J.Sm.
Coelogyne verrucosa S.E.C.Sierra
Coelogyne viscosa Rchb.f.
Coelogyne xyrekes Ridl.
Coelogyne yiii Schuit. & de Vogel
Coelogyne zurowetzii Carr

MEXICO / MEXIQUE (LE) / MÉXICO

Comparettia falcata Poepp. & Endl.
Masdevallia floribunda Lindl.

MYANMAR / MYANMAR (LE) / MYANMAR

Aerides crassifolia C.S.P.Parish ex Burb.
Acrides falcata Lindl & Paxton
Aerides flabellata Rolfe ex Downie
Aerides houlletiana Rchb.f.
Aerides multiflora Roxb.
Aerides odorata Lour.
Aerides rosea Lodd. ex Lindl. & Paxton
Aerides × jansonii Rolfe
Coelogyne assamica Linden & Rchb.f.
Coelogyne barbata Griff.
Coelogyne brachyptera Rchb.f.
Coelogyne calcicola Kerr
Coelogyne ecarinata C.Schweinf.
Coelogyne fimbriata Lindl.
Coelogyne flaccida Lindl.

167

Coelogyne fuliginosa Lodd. ex Hook.
Coelogyne fuscescens
Coelogyne griffithii Hook.f.
Coelogyne holochila P.F.Hunt & Summerh.
Coelogyne huettneriana Rchb.f.
Coelogyne lentiginosa Lindl.
Coelogyne leucantha W.W.Sm.
Coelogyne longifolia (Blume) Lindl.
Coelogyne longipes Lindl.
Coelogyne micrantha Lindl.
Coelogyne nervosa A.Rich.
Coelogyne nitida (Wall. ex D.Don) Lindl.
Coelogyne occultata Hook.f.
Coelogyne ovalis Lindl.
Coelogyne pallens Ridl.
Coelogyne parishii Hook.
Coelogyne picta Schltr.
Coelogyne prolifera Lindl.
Coelogyne pulchella Rolfe
Coelogyne punctulata Lindl.
Coelogyne rigida C.S.P.Parish & Rchb.f.
Coelogyne sanderae Kraenzl. ex O'Brien
Coelogyne schilleriana Rchb.f. & K.Koch
Coelogyne schultesii S.K.Jain & S.Das
Coelogyne stricta (D.Don) Schltr.
Coelogyne tenasserimensis Seidenf.
Coelogyne trinervis Lindl.
Coelogyne triplicatula Rchb.f.
Coelogyne ustulata C.S.P.Parish & Rchb.f.
Coelogyne viscosa Rchb.f.

NEPAL / NÉPAL / NEPAL

Aerides multiflora Roxb.
Aerides odorata Lour.
Coelogyne barbata Griff.
Coelogyne corymbosa Lindl.
Coelogyne cristata Lindl.
Coelogyne fimbriata Lindl.
Coelogyne flaccida Lindl.
Coelogyne fuscescens
Coelogyne holochila P.F.Hunt & Summerh.
Coelogyne longipes Lindl.
Coelogyne nitida (Wall. ex D.Don) Lindl.
Coelogyne ovalis Lindl.
Coelogyne prolifera Lindl.
Coelogyne punctulata Lindl.
Coelogyne raizadae S.K.Jain & S.Das
Coelogyne schultesii S.K.Jain & S.Das
Coelogyne stricta (D.Don) Schltr.

NEW CALEDONIA, OVERSEAS TERRITORY OF FRANCE /
NOUVELLE-CALÉDONIE, TERRITOIRE D'OUTRE-MER DE LA
FRANCE / NUEVA CALEDONIA, TERRITORIO DE ULTRAMAR DE
FRANCIA

Coelogyne lycastoides F.Muell. & Kraenzl.

NICARAGUA / NICARAGUA (LE) / NICARAGUA

Masdevallia chontalensis Rchb.f.
Masdevallia molossoides Kraenzl.
Masdevallia nicaraguae Luer
Masdevallia nidifica Rchb.f.
Masdevallia tubuliflora Ames

COMMONWEALTH OF THE NORTHERN MARIANA ISLANDS,
DEPENDANT TERRITORY OF THE UNITED STATES OF AMERICA /
COMMONWEALTH DE MARIANNES DU NORD, TERRITOIRE
DÉPENDANT DES ETATS-UNIS D'AMÉRIQUE / COMMONWEALTH
DE LAS ISLAS MARIANAS DEL NORTE, TERRITORIO
DEPENDIENTE DE LOS ESTADOS UNIDOS DE AMÉRICA

Coelogyne guamensis Ames

PALAU / PALAOS / PALAU

Coelogyne guamensis Ames

PANAMA / PANAMA (LE) / PANAMÁ

Masdevallia attenuata Rchb.f.
Masdevallia chontalensis Rchb.f.
Masdevallia collina L.O.Williams
Masdevallia eburnea Luer & Maduro
Masdevallia flaveola Rchb.f.
Masdevallia gloriae Luer & Maduro
Masdevallia lata Rchb.f.
Masdevallia livingstoneana Roezl
Masdevallia maduroi Luer
Masdevallia molossoides Kraenzl.
Masdevallia nidifica Rchb.f.
Masdevallia ostaurina Luer & V.N.M.Rao
Masdevallia picturata Rchb.f.
Masdevallia pleurothalloides Luer
Masdevallia rafaeliana Luer
Masdevallia scabrilinguis Luer
Masdevallia schizopetala Kraenzl.
Masdevallia striatella Rchb.f.
Masdevallia thienii Dodson
Masdevallia tokachiorum Luer
Masdevallia tonduzii Woolward

Part III: Country Checklist / Liste par Pays / Lista por Paises

Masdevallia utriculata Luer
Masdevallia zahlbruckneri Kraenzl.

PAPUA NEW GUINEA / PAPOUASIE-NOUVELLE-GUINÉE (LA) / PAPUA NUEVA GUINEA

Aerides quinquevulnera Lindl.
Coelogyne asperata Lindl.
Coelogyne beccarii Rchb.f.
Coelogyne carinata Rolfe
Coelogyne fragrans Schltr.
Coelogyne susanae P.J.Cribb & B.A.Lewis
Coelogyne veitchii Rolfe

PERU / PÉROU / PERÚ

Comparettia coccinea Lindl.
Comparettia falcata Poepp. & Endl.
Comparettia ignea P.Ortíz
Masdevallia abbreviata Rchb.f.
Masdevallia albella Luer & Teague
Masdevallia amabilis Rchb.f. & Warsz.
Masdevallia amplexa Luer
Masdevallia andreettaeana Luer
Masdevallia anomala Luer & Sijm
Masdevallia antonii Königer
Masdevallia aphanes Königer
Masdevallia ariasii Luer
Masdevallia asterotricha Königer
Masdevallia atahualpa Luer
Masdevallia audax Königer
Masdevallia aurorae Luer & M.W.Chase
Masdevallia ayabacana Luer
Masdevallia bangii Schltr.
Masdevallia barlaeana Rchb.f.
Masdevallia bennettii Luer
Masdevallia bicolor Poepp. & Endl.
Masdevallia bonplandii Rchb.f.
Masdevallia bryophila Luer
Masdevallia caloptera Rchb.f.
Masdevallia calosiphon Luer
Masdevallia campyloglossa Rchb.f.
Masdevallia cardiantha Königer
Masdevallia carnosa Königer
Masdevallia carpishica Luer & Cloes
Masdevallia castor Luer & Cloes
Masdevallia catapheres Königer
Masdevallia cinnamomea Rchb.f.
Masdevallia civilis Rchb.f. & Warsz.
Masdevallia cleistogama Luer
Masdevallia cloesii Luer
Masdevallia collantesii D.E.Benn. & Christenson

Masdevallia colossus Luer
Masdevallia concinna Königer
Masdevallia constricta Poepp. & Endl.
Masdevallia cordeliana Luer
Masdevallia coriacea Lindl.
Masdevallia cosmia Königer
Masdevallia cranion Luer
Masdevallia cuprea Lindl.
Masdevallia cyclotega Königer
Masdevallia davisii Rchb.f.
Masdevallia decumana Königer
Masdevallia dudleyi Luer
Masdevallia echo Luer
Masdevallia elegans Luer & R.Escobar
Masdevallia empusa Luer
Masdevallia ephelota Luer & Cloes
Masdevallia eumeces Luer
Masdevallia eumeliae Luer
Masdevallia figueroae Luer
Masdevallia formosa Luer & Cloes
Masdevallia fuchsii Luer
Masdevallia glandulosa Königer
Masdevallia goliath Luer & Andreetta
Masdevallia harlequina Luer
Masdevallia hymenantha Rchb.f.
Masdevallia icterina Königer
Masdevallia idae Luer & Arias
Masdevallia immensa Luer
Masdevallia instar Luer & Andreetta
Masdevallia ionocharis Rchb.f.
Masdevallia jarae Luer
Masdevallia juan-albertoi Luer & M.Arias
Masdevallia karineae Nauray ex Luer
Masdevallia klabochiorum Rchb.f.
Masdevallia kuhniorum Luer
Masdevallia lamprotyria Königer
Masdevallia leonii D.E.Benn. & Christenson
Masdevallia leptoura Luer
Masdevallia lilacina Königer
Masdevallia lilianae Luer
Masdevallia lineolata Königer
Masdevallia lintricula Königer
Masdevallia lucernula Königer
Masdevallia lychniphora Königer
Masdevallia manoloi Luer & M.Arias
Masdevallia marizae Luer & Rolando
Masdevallia melanopus Rchb.f.
Masdevallia mezae Luer
Masdevallia microptera Luer & Würstle
Masdevallia minuta Lindl.
Masdevallia monogona Königer
Masdevallia nikoleana Luer & J.Portilla
Masdevallia norops Luer & Andreetta
Masdevallia ortalis Luer

Masdevallia os-viperae Luer & Andreetta
Masdevallia oxapampaensis D.E.Benn. & Christenson
Masdevallia pandurilabia C.Schweinf.
Masdevallia paquishae Luer & Hirtz
Masdevallia parvula Schltr.
Masdevallia pernix Königer
Masdevallia phasmatodes Königer
Masdevallia phlogina Luer
Masdevallia phoenix Luer
Masdevallia picea Luer
Masdevallia picta Luer
Masdevallia picturata Rchb.f.
Masdevallia plantaginea (Poepp. & Endl.) Cogn.
Masdevallia plynophora Luer
Masdevallia pollux Luer & Cloes
Masdevallia polysticta Rchb.f.
Masdevallia popowiana Königer & J.G.Weinm.bis
Masdevallia posadae Luer & R.Escobar
Masdevallia pozoi Königer
Masdevallia princeps Luer
Masdevallia prodigiosa Königer
Masdevallia prolixa Luer
Masdevallia prosartema Königer
Masdevallia pumila Poepp. & Endl.
Masdevallia pyknosepala Luer & Cloes
Masdevallia pyxis Luer
Masdevallia recurvata Luer & Dalstrom
Masdevallia regina Luer
Masdevallia replicata Königer
Masdevallia rhodehameliana Luer
Masdevallia richardsoniana Luer
Masdevallia rigens Luer
Masdevallia rimarima-alba Luer
Masdevallia rodolfoi (Braas) Luer
Masdevallia rolandorum Luer & Sijm
Masdevallia rubeola Luer & R.Vasquez
Masdevallia rubiginosa Königer
Masdevallia rufescens Königer
Masdevallia rugulosa Königer
Masdevallia schizostigma Luer
Masdevallia schoonenii Luer
Masdevallia schroederiana Sander ex Veitch
Masdevallia scitula Königer
Masdevallia selenites Königer
Masdevallia semiteres Luer & R.Escobar
Masdevallia setacea Luer & Malo
Masdevallia spilantha Königer
Masdevallia stumpflei Braas
Masdevallia sulphurella Königer
Masdevallia terborchii Luer
Masdevallia titan Luer
Masdevallia tricallosa Königer
Masdevallia trochilus Linden & André
Masdevallia tubulosa Lindl.

Masdevallia uniflora Ruiz & Pav.
Masdevallia ustulata Luer
Masdevallia vargasii C.Schweinf.
Masdevallia veitchiana Rchb.f.
Masdevallia venusta Schltr.
Masdevallia vexillifera Luer
Masdevallia vomeris Luer
Masdevallia weberbaueri Schltr.
Masdevallia welischii Luer
Masdevallia wendlandiana Rchb.f.
Masdevallia whiteana Luer
Masdevallia wurdackii C.Schweinf.
Masdevallia × splendida Rchb.f.
Masdevallia xanthodactyla Rchb.f.
Masdevallia zebracea Luer

PHILIPPINES (THE) / PHILIPPINES (LES) / FILIPINAS

Aerides augustiana Rolfe
Aerides lawrenceae Rchb.f.
Aerides leeana Rchb.f.
Aerides odorata Lour.
Aerides quinquevulnera Lindl.
Aerides thibautiana Rchb.f.
Coelogyne asperata Lindl.
Coelogyne bilamellata Lindl.
Coelogyne candoonensis Ames
Coelogyne chloroptera Rchb.f.
Coelogyne confusa Ames
Coelogyne elmeri Ames
Coelogyne integerrima Ames
Coelogyne integra Schltr.
Coelogyne loheri Rolfe
Coelogyne longirachis Ames
Coelogyne marmorata Rchb.f.
Coelogyne merrillii Ames
Coelogyne palawanensis Ames
Coelogyne quinquelamellata Ames
Coelogyne remediosiae Ames & Quisumb
Coelogyne rochussenii de Vriese
Coelogyne sparsa Rchb.f.
Coelogyne swaniana Rolfe
Coelogyne usitana Roeth & O.Gruss
Coelogyne vanoverberghii Ames

PUERTO RICO, DEPENDANT TERRITORY OF THE UNITED STATES OF AMERICA / PORTO RICO, TERRITOIRE DÉPENDANT DES ETATS-UNIS D'AMÉRIQUE / PUERTO RICO, TERRITORIO DEPENDIENTE DE LOS ESTADOS UNIDOS DE AMÉRICA

Comparettia falcata Poepp. & Endl.

Part III: Country Checklist / Liste par Pays / Lista por Paises

SAMOA / SAMOA (LE) / SAMOA

Coelogyne lycastoides F.Muell. & Kraenzl.

SINGAPORE / SINGAPOUR / SINGAPUR

Coelogyne cumingii Lindl.
Coelogyne flexuosa Rolfe
Coelogyne foerstermannii Rchb.f.
Coelogyne testacea Lindl.

SOLOMON ISLANDS / ILES SALOMON / ISLAS SALOMÓN

Coelogyne asperata Lindl.
Coelogyne beccarii Rchb.f.
Coelogyne carinata Rolfe
Coelogyne susanae P.J.Cribb & B.A.Lewis
Coelogyne veitchii Rolfe

SRI LANKA / SRI LANKA / SRI LANKA

Aerides ringens (Lindl.) C.E.C.Fisch.
Coelogyne breviscapa Lindl.
Coelogyne odoratissima Lindl.
Coelogyne zeylanica Hook.f.

SURINAME / SURINAME (LE) / SURINAME

Masdevallia cuprea Lindl.
Masdevallia minuta Lindl.

THAILAND / THAÏLANDE (LA) / TAILANDIA

Aerides crassifolia C.S.P.Parish ex Burb.
Aerides falcata Lindl. & Paxton
Aerides flabellata Rolfe ex Downie
Aerides houlletiana Rchb.f.
Aerides krabiensis Seidenf.
Aerides multiflora Roxb.
Aerides odorata Lour.
Aerides rosea Lodd. ex Lindl. & Paxton
Coelogyne assamica Linden & Rchb.f.
Coelogyne calcicola Kerr
Coelogyne cristata Lindl.
Coelogyne cumingii Lindl.
Coelogyne fimbriata Lindl.
Coelogyne flaccida Lindl.
Coelogyne fuscescens
Coelogyne huettneriana Rchb.f.
Coelogyne lentiginosa Lindl.

Coelogyne longipes Lindl.
Coelogyne nitida (Wall. ex D.Don) Lindl.
Coelogyne ovalis Lindl.
Coelogyne pallens Ridl.
Coelogyne pulverula Teijsm. & Binn.
Coelogyne quadratiloba Gagnep.
Coelogyne radicosa Ridl.
Coelogyne rigida C.S.P.Parish & Rchb.f.
Coelogyne rochussenii de Vriese
Coelogyne schilleriana Rchb.f. & K.Koch
Coelogyne schultesii S.K.Jain & S.Das
Coelogyne septemcostata J.J.Sm.
Coelogyne tenasserimensis Seidenf.
Coelogyne tomentosa Lindl.
Coelogyne trinervis Lindl.
Coelogyne velutina de Vogel
Coelogyne virescens Rolfe
Coelogyne viscosa Rchb.f.
Coelogyne xyrekes Ridl.

TONGA / TONGA (LES) /TONGA

Coelogyne lycastoides F.Muell. & Kraenzl.

UNITED STATES OF AMERICA (THE) / ETATS-UNIS D'AMÉRIQUE (LES) / ESTADOS UNIDOS DE AMÉRICA (LOS)

Comparettia falcata Poepp. & Endl.

VANUATU / VANUATU / VANUATU

Coelogyne lycastoides F.Muell. & Kraenzl.
Coelogyne macdonaldii F.Muell. & Kraenzl.

VENEZUELA / VENEZUELA (LE) / VENEZUELA

Masdevallia amanda Rchb.f. & Warsz.
Masdevallia bicolor Poepp. & Endl.
Masdevallia caudata Lindl.
Masdevallia chontalensis Rchb.f.
Masdevallia clandestina Luer & R.Escobar
Masdevallia deceptrix Luer & Würstle
Masdevallia dunstervillei Luer
Masdevallia ensata Rchb.f.
Masdevallia garciae Luer
Masdevallia guayanensis Lindl. ex Benth.
Masdevallia impostor Luer & R.Escobar
Masdevallia irapana H.R.Sweet
Masdevallia iris Luer & R.Escobar
Masdevallia kyphonantha H.R.Sweet
Masdevallia lansbergii Rchb.f.

Masdevallia macroglossa Rchb.f.
Masdevallia maculata Klotzsch & H.Karst.
Masdevallia melanoxantha Linden & Rchb.f.
Masdevallia minuta Lindl.
Masdevallia navicularis Garay & Dunst.
Masdevallia norae Luer
Masdevallia pachysepala (Rchb.f.) Luer
Masdevallia picturata Rchb.f.
Masdevallia rechingeriana Kraenzl.
Masdevallia sanctae-fidei Kraenzl.
Masdevallia sceptrum Rchb.f.
Masdevallia schlimii Linden ex Lindl.
Masdevallia sprucei Rchb.f.
Masdevallia stirpis Luer
Masdevallia striatella Rchb.f.
Masdevallia strumifera Rchb.f.
Masdevallia tovarensis Rchb.f.
Masdevallia triangularis Lindl.
Masdevallia tubulosa Lindl.
Masdevallia venezuelana H.R.Sweet
Masdevallia verecunda Luer
Masdevallia wageneriana Linden ex Lindl.
Masdevallia wendlandiana Rchb.f.
Masdevallia × synthesis Luer
Masdevallia × wubbenii Luer

VIET NAM / VIET NAM (LE) / VIET NAM

Aerides crassifolia C.S.P.Parish ex Burb.
Aerides falcata Lindl. & Paxton
Aerides houlletiana Rchb.f.
Aerides multiflora Roxb.
Aerides odorata Lour.
Aerides rosea Lodd. ex Lindl. & Paxton
Aerides rubescens (Rolfe) Schltr.
Coelogyne assamica Linden & Rchb.f.
Coelogyne brachyptera Rchb.f.
Coelogyne calcicola Kerr
Coelogyne dichroantha Gagnep.
Coelogyne eberhardtii Gagnep.
Coelogyne filipeda Gagnep.
Coelogyne fimbriata Lindl.
Coelogyne flaccida Lindl.
Coelogyne fuscescens
Coelogyne griffithii Hook.f.
Coelogyne huettneriana Rchb.f.
Coelogyne lawrenceana Rolfe
Coelogyne lentiginosa Lindl.
Coelogyne lockii Aver.
Coelogyne malipoensis Z.H.Tsi
Coelogyne mooreana Rolfe
Coelogyne ovalis Lindl.
Coelogyne pallens Ridl.

176

Coelogyne quadratiloba Gagnep.
Coelogyne rigida C.S.P.Parish & Rchb.f.
Coelogyne sanderae Kraenzl. ex O'Brien
Coelogyne schultesii S.K.Jain & S.Das
Coelogyne stricta (D.Don) Schltr.
Coelogyne tenasserimensis Seidenf.
Coelogyne trinervis Lindl.
Coelogyne virescens Rolfe
Coelogyne viscosa Rchb.f.

UNKNOWN

Masdevallia fosterae Luer
Masdevallia × polita Luer & Si